AF411790

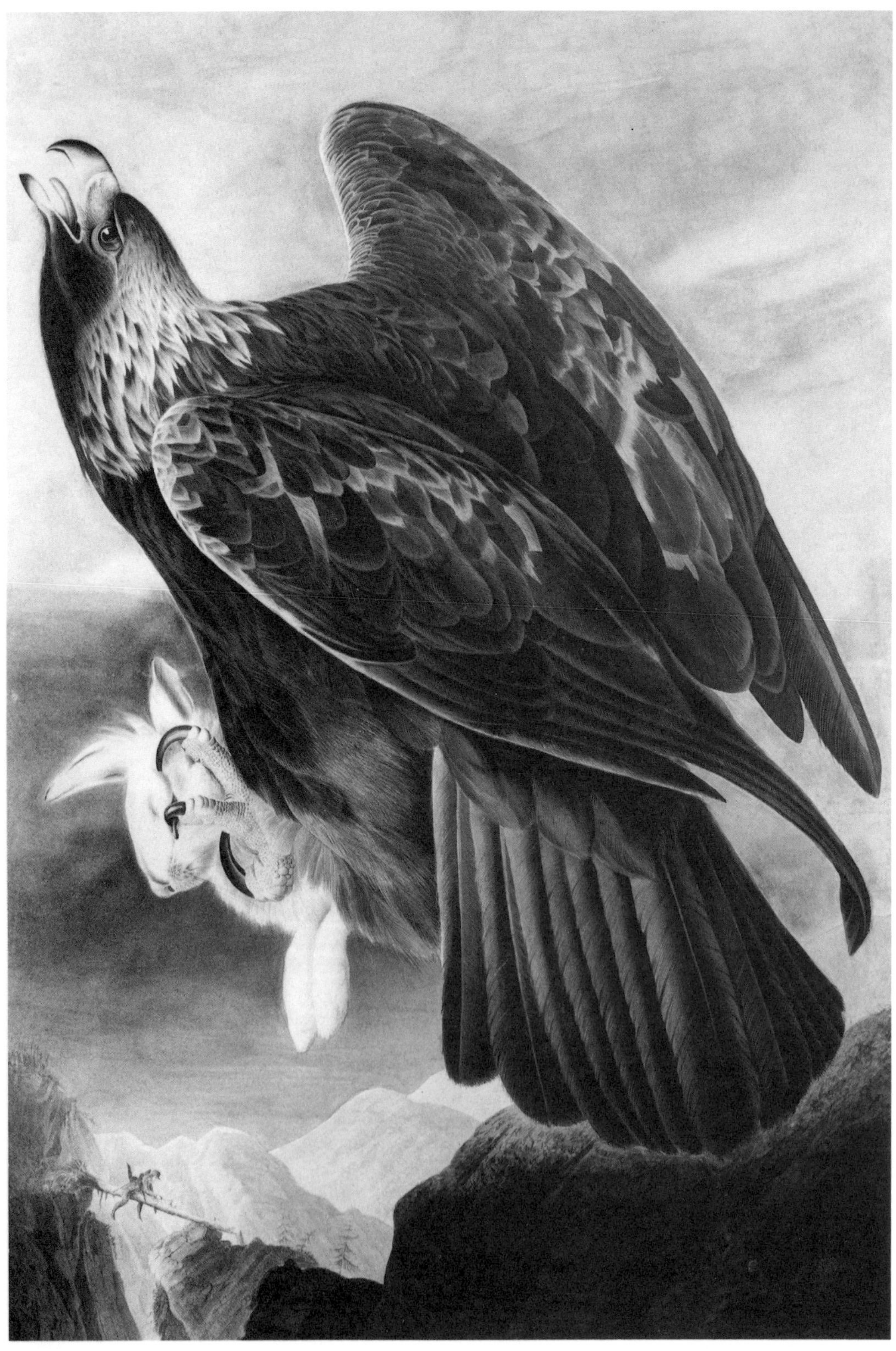

John James Audubon, *The Golden Eagle* (1833); watercolor, 95.4 x 64.7 cm. Collection of the New-York Historical Society.

Art and Science
in America

Art and Science in America

Issues of Representation

edited by Amy R. W. Meyers

Huntington Library
San Marino, California

Cover illustration

Orchard oriole, from Alexander Wilson, *American Ornithology*, volume 1 (1808).
Huntington Library.

The essays in this volume are based on presentations given at a symposium at the Huntington,
"Art and Science in America: Issues of Representation," held on 18–19 March 1994.

Cataloging-in-Publication Data

Art and science: issues of representation / edited by Amy R. W. Meyers.
    p. cm.
"The essays in this volume are based on presentations given at a symposium at the Huntington,
'Art and Science in America: Issues of Representation,' held on 18–19 March 1994"—T.p. verso.

Includes bibliographical references.
ISBN 0-87328-172-1

1. Natural history museums—United States—Congresses. 2. Botanical gardens—United States—Congresses.
3. Ornithological illustration—United States—Congresses. 4. Landscape painting—United States—
Congresses. 5. Naturalists—United States—Biography—Congresses. 6. Landscape painters—United States—
Biography—Congresses. I. Meyers, Amy R. W. (Amy R. Weinstein). II. Art and Science in America: Issues of
Representation (1994: Huntington Library)

QH70.U6A78 1998   98–14245
508—dc21

Contents

Introduction

AMY R. W. MEYERS

In recent years, the Huntington Library, Art Collections, and Botanical Gardens has become a center for the study of the visual and material culture of science. Scholars trained in a variety of disciplines have been drawn to the institution's rich collections relating to the history of scientific research and exploration in order to investigate the ways in which two-dimensional representations of the natural world (paintings, drawings, prints, and maps), as well as illustrated books, cabinets of preserved specimens, botanical gardens, menageries, and museum displays, all contributed to the development of Western cosmology, from the Renaissance through the first decades of the twentieth century.

In 1993, Robert C. Ritchie, the Huntington's Director of Research, and Edward J. Nygren, Director of the Art Division, agreed that a symposium should be organized to highlight the importance of the Huntington's holdings to this newly emerging interdisciplinary undertaking. Since particular attention has been paid to materials pertaining to the visual and material culture of American science of the eighteenth and nineteenth centuries, this general subject was selected for the conference. As the Huntington's Curator of American Art, with a special interest in the the relationship between art and science, I was approached to coordinate the proceedings, which were scheduled for March of the following year. The invitation provided a welcome opportunity to survey the state of this rapidly growing field and to formulate sessions that might represent some of the most prominent avenues of inquiry. I was joined in this endeavor by David R. Brigham, then Research Associate in American Art, who, through his work on the community of interest that supported Charles Willson Peale's museum in Philadelphia during the post-Revolutionary period, was able to offer invaluable suggestions for session topics and participants.

Wishing to explore a wide array of methodological approaches but realizing that we would need to provide coherence among the presentations, we narrowed the time frame to cover the years spanning the establishment of the nation through the first decades of the Republic. We divided the conference into three sessions, each designed to examine a particular type of artistic activity from a different analytical point of view. In the first session, "Arranging the Stuff of Nature: The Botanic Garden and the Natural History Museum," David Brigham began the proceedings by examining Peale's didactic program for exhibiting to the citizenry of the new nation the principles of a natural social order. Through an analysis of Peale's approach to the amassing of his collections, as well as his preparation and presentation of them, David illuminated the political and social objectives that Peale invested in his museum, as a vehicle to promote the proper model for human relationships derived from patterns of interaction among animals and plants, ordered according to the ancient concept of the Great Chain of Being. In the same session of the conference, Therese O'Malley explored the history of American botanical gardens from the Revolution through the antebellum period, grounding an assessment of the scientific garden as a "national object" in Peale's attempts to fashion his politically and socially meaningful garden at Belfield. Her discussion came to focus particularly on an interpretation of the National Mall in Washington, D.C., as a botanic garden, reshaped over time to fit the changing symbolic needs of a rapidly expanding nation.

The examination of Peale's activities as a case study in assessing the iconographic and didactic power of museum display for the purposes of promoting the proper social order was also undertaken by Laura Rigal. Her appraisal of Alexander Wilson's *American Ornithology*, delivered in the session "Of Birds and Men: The Cultural Import of Ornithological Illustration," placed Wilson, as the producer of the first illustrated ornithological work to be published in America, within Peale's community; and it argued for the inherent character of both men's work as Federalist projects. Rigal suggested that Wilson's *Ornithology*, like Peale's Museum, was the monumental outcome of one man's vision, joining "labor with looking." Reflecting Wilson's herculean task of uniting huge numbers of local specimens within a national collection, his *Ornithology* also called on the reader to exert the energy necessary to comprehend the colossal illustrated production—a task of visual and textual analysis posited by Wilson as equivalent to the independent labors of Jefferson's "virtuous farmer."

The self-conscious role of the naturalist-artist as an industrious chronicler of information was also considered, in the same session, by Linda Dugan Partridge, who addressed the myth that Audubon perpetuated about his own exercises as a "self-taught woodsman who followed only nature and his genius in painting American birds." Partridge explored the many ways in which the naturalist in fact relied on older European publications and conventions of illustration to create pictures that he claimed to represent firsthand empirical observations. In the course of her argument, she provided a critique of nationalist positivism, with its emphasis on progress and the rejection of the old, and she pointed to the ways in which traditional conventions actually governed new productions.

Questions concerning the fundamental character of the visual analysis of nature in America occupied Kenneth Haltman as well. Delivered in the session "Depicting Deep Time: Early Geological Illustration," his paper examined landscapes by Samuel Seymour executed on the Stephen H. Long expedition (1819–20)—the first scientific survey of the western frontier to be sponsored by the federal government. Haltman traced the dialectic found in Seymour's images, between their function as objective descriptions of geologic structures, driven by the positivist thrust of science, and their more self-referential role as commentaries on artistic representation as the product of imagination. Indeed, he argued that Seymour saw his drawings as a means to interpret not only the nature of art but also the essential character of science as an activity engaged in by contemporary Americans to penetrate the meaning of physical creation and to possess its bounties.

Whereas Haltman addressed the work of an artist who, through his drawings, grappled directly with the significance of the most advanced geologic theories of his day, Rebecca Bedell examined the work of an artist, Thomas Cole, who remained rooted in a conservative view of geology as bound to revealed religion. Bedell argued that Cole did not conceive of geology as the subject of his work but as a vehicle by which to reinforce the moral and spiritual teachings that he sought to convey through his paintings. She also proposed that Cole's study of the science, as well as his use of careful geologic description in some of his paintings, allowed him to align himself with a group of powerful patrons of the gentry class who were, themselves, immersed in the study of geology from the same conservative, theological perspective.

Whether the relationship between art and science was seen to involve issues of social standing, religious faith, political imperative, economic interest, aesthetic principles, or philosophical concerns was charted clearly in each symposium paper by the session moderators, Bruce Robertson and Robert Ritchie. They brought to their commentaries a depth of knowledge about the portrayal of landscape and nature in eighteenth- and nineteenth-century America that enabled them to place the themes explored by the participants in the broadest of cultural contexts. Michael Kammen, Times Mirror Foundation Visiting Chair of American Studies at the Huntington for the year, also extended the parameters of the discussion concerning the relationship between empirical observation and visual portrayal in an informal talk entitled "The Vision Thing: The Role of Spectacles in American Art and Culture."

Susan Green, the editor of the *Huntington Library Quarterly*, noted the value of the papers in establishing a benchmark in the historical analysis of the interplay between the visual arts and the natural sciences in America; and we soon began to make plans for a publication based on the proceedings. Providing her own astute commentary in the course of the editing process, Susan encouraged the authors to sharpen their ideas, amplify their arguments, and refine their prose. Her aid to me in coordinating the publication was indispensable, and she served as the guiding force bringing the work to completion. I am also indebted to Kathleen Thorne-Thomsen for the elegant design of the volume, and to Carolyn Powell, Ariel Presta, and Michelle Kaufman for their help with the conference proceedings. To Robert Ritchie and Edward Nygren I would like to extend my gratitude for their continuing belief in the value of the publication, and for their support in seeing it come to light.

List of Figures

Sources and acknowledgments appear in the captions.

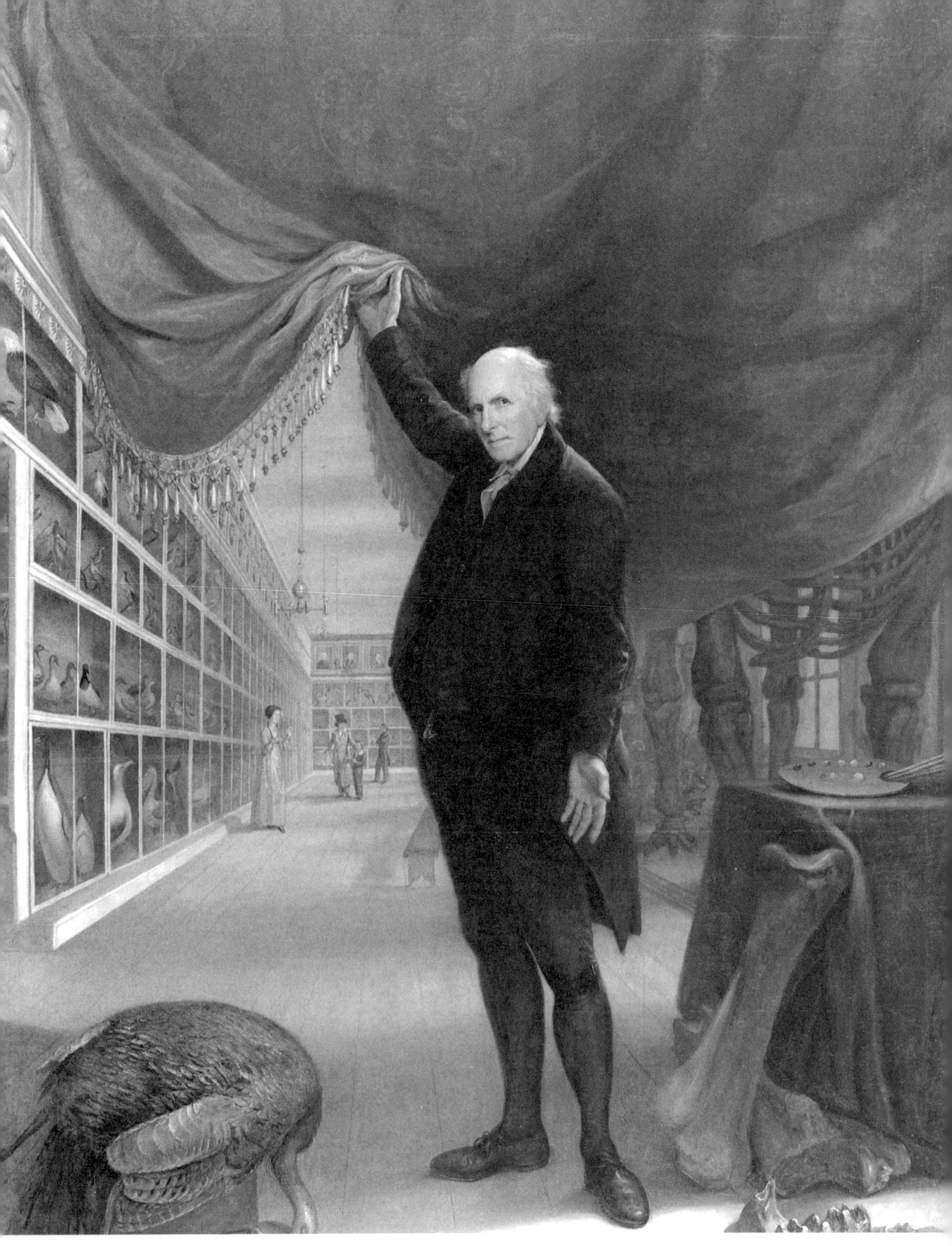

Figure 1. Charles Willson Peale, *The Artist in His Museum* (1822); oil on canvas, 261.6 x 202.7 cm. Pennsylvania Academy of the Fine Arts, Philadelphia, gift of Mrs. Sarah Harrison.

"Ask the Beasts, and They Shall Teach Thee": The Human Lessons of Charles Willson Peale's Natural History Displays

Davᴅ R. Brigham

Through the display and interpretation of natural history specimens, Charles Willson Peale expressed his belief that animal behavior provided a model for an economically productive, socially harmonious, and morally upright republic. From the distance of two centuries, we recognize such a model as thoroughly cultural—that is, constructed by humans to serve human interests. Peale transformed nature for cultural purposes in a number of different ways and with a variety of tools, including the rifle, the scalpel, the needle, the paintbrush, and the pen. He hunted, gutted, and stuffed animals, and arranged these specimens in lifelike poses; he then painted habitat settings in which to display them; and he wrote labels, catalogues, and lectures to interpret them. One of Peale's central ideas was that nature is simultaneously hierarchical and harmonious, and by means of his displays he extended this construction of nature to human life. A republic that was at once rank-ordered and cooperative would be optimally satisfying, abundant, and wholesome. Peale and his followers promoted this social and political set of ideas not only as natural but also as the expression of God's will.

Collecting Specimens

When Charles Willson Peale opened his "Repository for Natural Curiosities" in 1786, he was adding nature as a category of display to that of art.[1] Up until this time, Peale had exhibited his portraits in oil of Revolutionary War heroes, transparent paintings of local landscapes and scenes from Milton's *Paradise*

---

1. See *Pennsylvania Packet,* 18 July 1786.

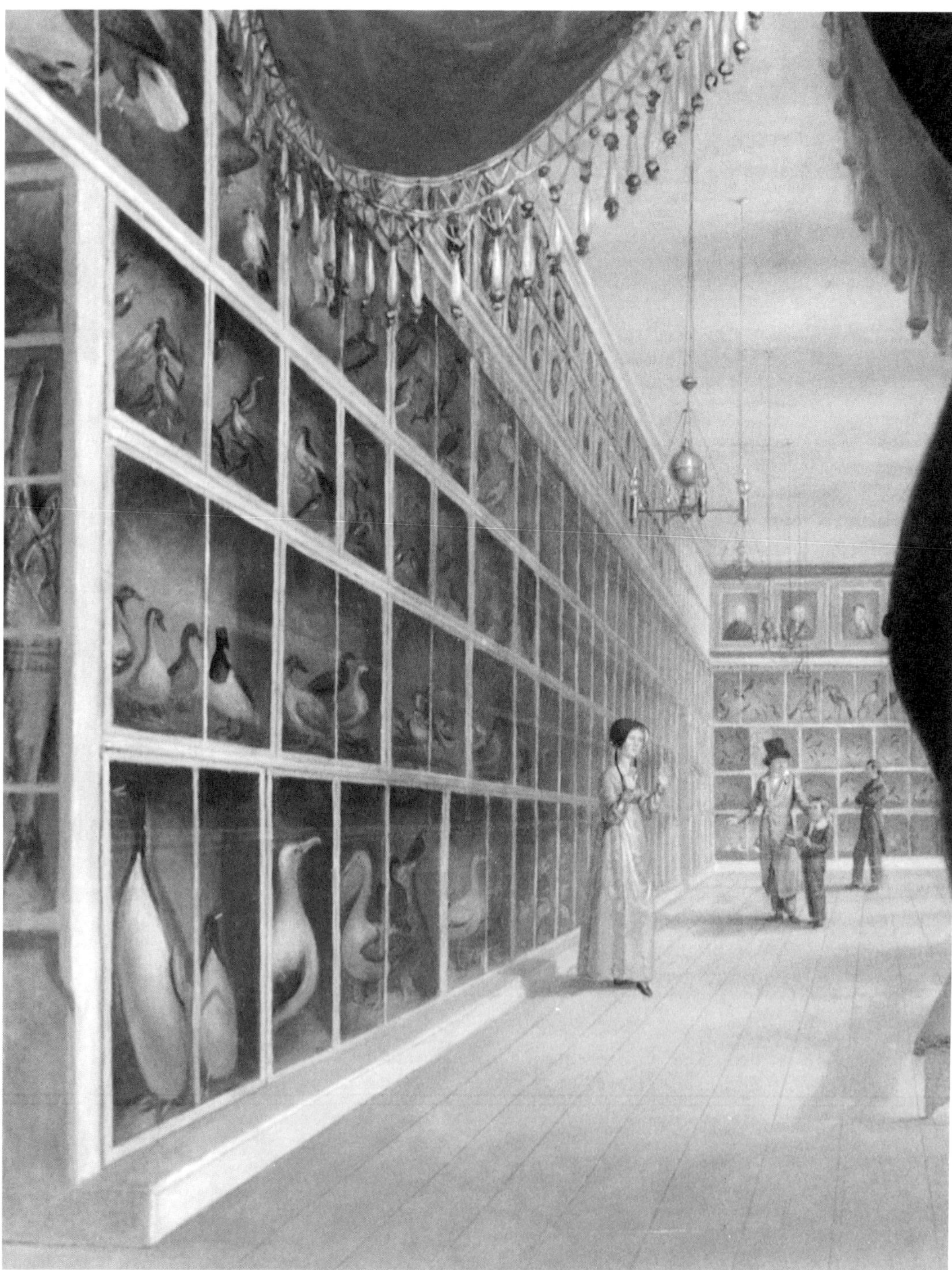

Figure 2. Peale, *The Artist in His Museum* (detail).

Lost.[2] In *The Artist in His Museum,* Peale's late self-portrait, he depicts the repository in 1822, almost forty years after its opening. The long exhibition hall is shown, its cases filled with preserved bird specimens arranged in an orderly gridlike pattern (fig. 1). Contemporary guidebooks state that Peale had accumulated about eleven hundred bird specimens—along with hundreds of quadrupeds, fish, reptiles, insects, minerals, and shells—by the time this portrait was painted. Peale himself estimated that the collections contained over eighteen hundred birds and more than a hundred thousand objects total by this time.[3] Peale developed his collections by two means: by soliciting and receiving donations of specimens; and by hunting birds and other animals.

As shown in the detail taken from the left margin of *The Artist in His Museum,* Peale acknowledges the importance of donations for the growth of his collections (fig. 2). The specimen featured in this elongated case is a paddlefish, and the inscription proclaims that it is the gift of Robert Patterson, a prominent professor at the University of Pennsylvania. Peale encouraged such contributions by publishing lists of recent acquisitions and the names of their donors in Philadelphia's newspapers.[4] Such notices also served, of course, to announce the expansion of his collection. Peale appealed to individual donors and recognized their assistance privately with printed gift acknowledgments on which an allegorical figure pulls back the curtain to reveal the wonders of nature, just as Peale does in his self-portrait (fig. 3). Through gifts to the museum, donors demonstrated their shared interest in educating citizens of the new republic; they, too, were capable of pulling back the veil of ignorance. The motto in the gift acknowledgment, "Of grains are Mountains form'd," indicates that Peale amassed his substantial collection from individual contributions—and suggests his belief in an analogy between his enterprise and natural processes.

The appeals for specimens were responded to by a wide range of individuals, from merchants who traded in foreign ports to members of exploration parties and eminent naturalists. When Robert Morris's ship *Alliance* returned from China in 1788, this wealthy merchant donated a silver pheasant

---

2. See, for example, *Independent Gazetteer,* 21 May 1785, where an advertisement describes a view of a country seat from night to dawn and in full daylight; a street at evening, then at night lit by street lamps; a "grand piece of architecture" in rain and then with rainbow; a representation of Milton's Pandemonium; and a mill near the falls of the Schuylkill. These scenes were transparent pictures with "changeable effects," modeled on the theatrical pictorial displays presented in London by Philip James de Loutherbourg.

3. Once such guidebook was *Philadelphia in 1824* (Philadelphia, 1824), 102–3; Peale's estimate is found in Charles Coleman Sellers, *Charles Willson Peale* (New York, 1969), 346.

4. For an enumeration of these lists, see David R. Brigham, "'A World in Miniature': Charles Willson Peale's Philadelphia Museum and Its Audience, 1786–1827" (Ph.D. diss., University of Pennsylvania, 1992), 358–429.

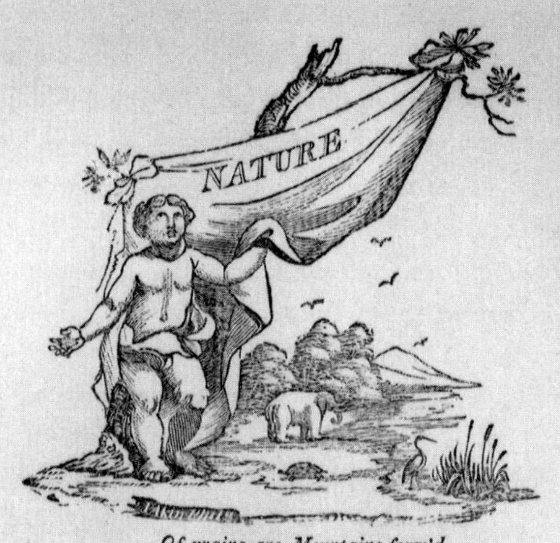

To advance the interest of the Museum, or Repository of Nature and Art, is indirectly conducing to the public benefit; permit me therefore, to join my assurances of obligation for any assistance you may render the institution by

And since it is from the science, zeal and liberality of individuals that we must be indebted for the developement of Nature's boundless stores, as well as the ingenuity of art; it shall be my endeavour so to dispose of them as may insure their preservation and public utility.

Sir, with respect,

Museum, Philadelphia,   180

Figure 3. Peale's gift acknowledgment. American Philosophical Society.

and exotic wares collected by the ship's captain in Asia.[5] Both privately and federally sponsored expeditions yielded significant additions to Peale's natural history collections, including plants and minerals gathered by Lewis and Clark, and quadrupeds, birds, and reptiles collected by the second Titian Ramsay Peale (1799–1885) during the Long Expedition.[6] Through exchanges of specimens, Peale also cultivated relationships with naturalists in America and abroad, such as ornithologists Alexander Wilson and John Latham, French mineralogist René-Just Haüy, British naturalist Sir Joseph Banks, and Thomas Jefferson.[7]

Peale also gathered specimens for his collection by hunting them himself. He frequently ventured into the woods and fields and on the river banks and beaches near Philadelphia, and he occasionally made trips as far as Maryland, Delaware, and New Jersey. In the late 1780s, the first years of the museum's operation, Peale's attention was still chiefly consumed by his business as a portraitist. Nonetheless, when he traveled by horse-drawn chaise through Maryland in 1789, painting watercolor-on-ivory miniatures and oil paintings for patrons there, he recorded in his diary that he had collected about sixty birds for the museum.[8] During the 1790s, Peale finally turned his attention fully to the business of building and managing the museum.[9] Peale's diary entries show that the trips to Maryland, Delaware, and New Jersey now focused on hunting animals for the displays.

5. Peale's diary, 23 November 1788, in *The Selected Papers of Charles Willson Peale and His Family*, ed. Lillian B. Miller et al. (New Haven, Conn., and London, 1983–), vol. 1, *Charles Willson Peale: Artist in Revolutionary America, 1735–1791*, 547.
6. The Long Expedition (1819–20), named after its commander Stephen H. Long, investigated the geology of the central plains and Rockies. On the Lewis and Clark accession, see "Memoranda of the Philadelphia Museum," in the Manuscripts Collection of the Historical Society of Pennsylvania, Philadelphia, entry for 28 December 1809, 43–45; on the Long Expedition, see "Invoice of Zoological Specimens and Drawings prepared by Titian Peale," "Memoranda," entry for 23 March 1821, 112–13. Peale named two sons Titian Ramsay Peale; the first lived 1780–98, the second 1799–1885. Both pursued their father's interest in art and natural history.
7. On Wilson's donations, see "Memoranda" (first section of Wilson's *American Ornithology*), 13 July 1807, 24; birds: 28 December 1810, 53; October 1811, 56; and 30 November 1812, 64; nests and eggs: 14 July 1808; and minerals: 30 December 1808, 37; and 6 June 1809, 41. On Latham's gift of birds from New Guinea and New Holland, see "Memoranda," 4 October 1807, 25. On Haüy's gifts of minerals, see "Memoranda," 30 December 1808, 38; and 12 March 1810, 49–50. On Banks's gift of a fossil shark's tooth, see "Memoranda," 1805, 11. On Jefferson's contributions, see "Memoranda," minerals: 10 May 1805, 5; and 28 December 1809, 45; quadrupeds: 17 October 1805, 8; and 28 January 1808, 28; and Native American artifacts: 28 December 1809, 43–45.
8. On Peale's painting trip to Maryland, see Peale's diary, 29 April–18 May 1789, in Miller, *Selected Papers,* 1:561–63.
9. Peale went so far as to announce his retirement from portrait painting; see *Dunlap and Claypoole's American Daily Advertiser,* 24 April 1794, in Miller, *Selected Papers,* vol. 2, *Charles Willson Peale: The Artist in His Museum, 1791–1810,* 91–92. However, he continued to paint throughout this period, and he resumed taking commissions; see Miller, *Selected Papers,* 2:240–41 n. 7.

Whereas records of donations distance the collector from the kill, Peale's diary makes concrete the role of the gun in gathering specimens. In 1793 Peale sailed for Cape Henlopen, Delaware, with his wife Elizabeth, infant son Vandyke, and the first Titian, then thirteen and already his museum assistant.[10] On 22 August, during a stop on Pea Patch Island, Peale "went ashore and shot a number of small Birds." Two days later, "a considerable number of Mother Carrys Chickens came round the Vessel and we shot 5 of them, we might have taken many more of them but this number were sufficient for my purpose." Titian, also skilled with his rifle, "shot a flood Gull & a White Crain & some Plover" on 27 August. Two days later, the father-and-son team borrowed a flat-bottom boat and hired two boys to row for them, and they "shot . . . willets, mud hens, Curlews, Gull &c." Peale coordinated trips to locations known to be rich in desirable species of birds. He was assisted by his sons Raphaelle and Titian, as well as by a slave, a wage laborer, and local youths who worked as day laborers. Together they hiked, rowed boats, and shot animals. It was thus not only a personal blow but also a significant professional setback when Titian died in 1798 at age eighteen.[11] He had been a promising naturalist and Peale's principal assistant both at the museum and on expeditions. Peale then had to find additional personnel. The diary records that on 30 May 1799, Peale sailed for Cape May, New Jersey, with Jotham Fenton and Moses Williams.[12] Fenton was a mechanic, skilled in taxidermy and other arts useful to a museum keeper, whom Peale had hired away from the New York Museum a few years earlier. Moses Williams was Peale's slave, referred to in his diary as "my Molatto Man Moses."

That hunting was the standard way to gather specimens is indicated not only in Peale's private diary but also in many public contexts. For example, he requested in a 1787 advertisement: "Mr. Peale's respectful compliments to the gentlemen sportsmen of Philadelphia, and [he] will be obliged to them for such birds and Beasts as are not yet preserved in Mr. Peale's Museum."[13] Mark Catesby, in the preface to his *Natural History of Carolina*, had earlier acknowledged the importance of the hunt to the practice of natural history. With Native Americans as his guides, Catesby hunted birds, buffalo, bears, panthers, "and other wild Beasts."[14] Both Peale and Catesby described hunting activi-

---

10. On Peale's trip to Cape Henlopen, see Peale's diary, 21 August–19 September 1793, ibid., 2:50–67.
11. Ibid, 2:226, on Titian Peale's untimely death, 18 September 1798.
12. On Peale's collecting trip to Cape May, see Peale's diary, 30 May–12 June 1799, ibid., 2:241–42.
13. *Pennsylvania Packet,* 11 September 1787.
14. Mark Catesby, *The Natural History of Carolina, Florida and the Bahama Islands* (London, 1731–43 [1729–47]), preface, 1:viii. Catesby also records his inability to shoot a yellow-breasted chat, and he finally had to "employ an Indian, who did it not without the utmost of his skill" (plate 50).

ties in a matter-of-fact tone, and they were apparently untroubled by the contradiction in killing specimens to display or depict them as though living. The Romantic artist and naturalist, however, developed an uneasiness about shooting that had not troubled their Enlightenment predecessors. For example, John James Audubon's writings reveal his ambivalence about killing the creatures he loved, but he also strove to romanticize his role as collector, characterizing himself as "the American woodsman."[15] In his watercolor *The Golden Eagle* (frontispiece), he presents himself (dressed in buckskin, precariously straddling a tree that bridges a rocky gulf) as a predator who, like the eagle who must kill to survive, commits violence for the "sustenance" of science and art.

Taxidermy

A vital concern among eighteenth-century naturalists was the development of a method to preserve specimens from rotting and from destruction by insects. Collectors insisted, further, that a successful mount should recreate the appearance and characteristics of a living animal. Some scientific writers, in the language they used to describe the reanimation of specimens, even conflated human and animal qualities. Peale fully embraced the anthropocentric approach to taxidermy, and he employed it to advance the museum's lessons that republican economic, social, and political life ought to follow the example of nature.

Peale's diaries from his collecting trips demonstrate that he hunted and preserved specimens nearly simultaneously. During his 1793 trip to Cape Henlopen, he noted that, shortly after shooting five "Mother Carrys Chickens," "Titian & myself set to work and dressed two of the Birds."[16] Several days later, they began the day by gutting two birds. (Then they ate breakfast.) Titian later shot several birds, "the mounting of which keep us busey the remaining part of the day." Later in the course of the same trip, Peale sent Titian out hunting by himself and stayed behind to dissect and treat the birds that his son brought back each day.

15. My reading of Audubon derives from Amy R. W. Meyers, "Observations of an American Woodsman: John James Audubon as Field Naturalist," in Annette Blaugrund and Theodore E. Stebbins Jr., eds., *John James Audubon: The Watercolors for "The Birds of America"* (New York, 1993), 43–54; and from Linda Dugan Partridge, "Domestic Violence: Scientific Themes and Audubon's Rattlesnake Attacked by Mockingbirds" (paper presented at the annual meeting of the American Studies Association, 6 November 1992); and Partridge, "The Construction of the Artist-Naturalist: Audubonography" (paper presented at the annual meeting of the College Art Association, 4 February 1993).

16. See Peale's diary, 21 August–19 September 1793, in Miller, *Selected Papers*, 2:50–67. The Mother Carey's chicken is better known as a storm petrel.

In the eighteenth century, collectors often consulted manuals for advice on how to preserve the fish, reptiles, birds, and quadrupeds they hunted. Instructions were also published as broadsheets, in handbooks directed at would-be naturalists, in scientific journals, and in travel accounts.[17] Writers of such advice addressed themselves both to their peers at home and to correspondents far afield who had access to animals in Africa, Asia, and the Americas. Storing specimens in spirits was one method suggested to prevent rotting. These fluids served as effective preservatives but did not make it possible to regain the appearance of a living specimen, and therefore most naturalists preferred to gut their specimens. A manual written by Étienne François Turgot and published in Paris in 1758 advocated skinning an animal rather than immersing it in spirits, and the plates demonstrated the operation.[18] Others suggested the additional precaution of baking the carcasses at a low temperature to burn the fat off the skin gradually. To preserve their lifelike appearance, Peale, too, gutted his specimens, and he did so soon after killing them. Hunting on a creek with Titian, Peale recorded in his diary that their boat ran aground, leaving them temporarily stranded. "That we might save the birds from spoiling," Peale wrote, "I took the bowels & brains out, and put some bur[n]t allum in those places."[19]

Insects were a constant threat to collections. Peale complained to his audience in a 1792 newspaper address that "his labours herein have be[en] great, and disappointments many; especially respecting proper methods of preserving dead animals from the ravage of moths and worms."[20] Storage in spirits, while effective, destroyed the appearance of the specimen, as noted above. Baking, used to prevent rotting initially, could be repeated occasionally to destroy insects and their eggs. Naturalists also experimented with various chemical treatments of skins. Collectors were advised to sprinkle animals with alum and other salts, but these substances caused discoloration and deterioration of the skin. Tobacco, cinnamon, pepper, and a variety of strong spices were tried, but these proved ineffective. Finally, the use of poisons, including arsenic and sublimate of mercury, was proposed. Although Peale frequently

---

17. For a survey of such methods used in the eighteenth century, see Paul Lawrence Farber, "The Development of Taxidermy and the History of Ornithology," *Isis* 68 (December 1977): 550–66.

18. Étienne François Turgot, *Mémoire instructif sur la manière de rassembler, de preparer, de conserver, et d'envoyer les diverses curiosités d'histoire naturelle* (Paris, 1758), plate 3.

19. Peale's diary, 5 September 1793, in Miller, *Selected Papers*, 2:58. Peale also described his method in "Directions for Preserving Birds, &c.," which he apparently enclosed in a letter of 26 July 1787 to Ebenezer Hazard; see Miller, *Selected Papers*, 1:488–89.

20. "To the Citizens of the United States of America," *Dunlap's American Daily Advertiser*, 13 January 1792; Miller, *Selected Papers*, 2:9.

complained about the inadequacies of published methods, he typically treated the skin and feathers, or fur, with a solution containing arsenic.[21]

Once spoilage and infestation were averted, the next challenge was to reconstruct the specimen's lifelike appearance. In fact, this concern affected the prescribed manner of the kill as well. Tesser Samuel Kuckahn cautioned readers of the *Philosophical Transactions* to use gunshot of as small a gauge as possible to limit damage.[22] Similarly, John Reinhold Foster, in his *Catalogue of the Animals of North America,* advised collectors of quadrupeds to take care that "the hair of the fur [is] as little stained with blood as possible."[23] Various scientific publications, including taxidermy manuals, suggested methods for preserving a living appearance. In the *Philosophical Transactions,* René-Antoine Ferchault de Reaumur advised putting feathers "into their natural Order," to make the specimen "appear as it was when it was alive."[24] John Lettsom argued the importance of installing "eyes as near the natural ones as possible."[25] To reproduce the original form of the animal after gutting, specimens were stuffed with cotton, oakum (hemp compounded with tar), straw, hay, wool, flax, or other soft materials. Turgot's manual proposes the use of cotton to fill the cavities of the abdomen and head (fig. 4). Wire was recommended for holding legs, wings, and head in characteristic poses.[26] Peale stuffed small birds with cotton but filled large birds with oakum because of its greater stiffness. For quadrupeds, Peale drew upon his skill as a sculptor and carved forms in wood to replicate their limbs and musculature.[27]

Peale prided himself on his skill as a taxidermist, and he incorporated his ability to restore a vibrant appearance to a dead animal into the narrative of *The Artist in His Museum.*[28] In the left foreground, Peale included a limp

21. Peale's diary, 18 May 1790, ibid., 1:563.
22. Tesser Samuel Kuckahn, "Four Letters from Mr. T. S. Kuckhan [*sic*], to the President and Members of the Royal Society; On the Preservation of dead Birds," *Philosophical Transactions* 60 (1770): 305.
23. John Reinhold Foster, *A Catalogue of the Animals of North America . . . To which are added, Short Directions for Collecting, Preserving, and Transporting, All Kinds of Natural History Curiosities* (London, 1771), 35.
24. M. d Reaumur, "Divers Means for preserving from Corruption dead Birds, intended to be sent to remote Countries, so that they may arrive there in a good Condition," *Philosophical Transactions* 45, no. 487, (April–June 1748): 307.
25. John Coakley Lettson, *The Naturalist's and Traveller's Companion,* 3d ed. (London, 1799), 15.
26. See Turgot, *Mémoire instructif,* plate 6 (figure 4, below); see also Foster, *Catalogue of the Animals of North America,* 37.
27. Peale, "Directions for Preserving Birds, &c." (1787); in Miller, *Papers,* 1:983, 488–89; and Peale, "A Walk through the Philadelphia Museum" [1805–6], 7, in Lillian B. Miller, ed., *The Collected Papers of Charles Willson Peale and His Family,* Kraus Microform, IID/27.
28. See Roger B. Stein, "Charles Willson Peale's Expressive Design: *The Artist in His Museum,*" *Prospects* 6 (1981): 143–44, 160, and 168; and Laura Rigal, "An American Manufactory: Political Economy, Collectivity, and the Arts in Philadelphia, 1790–1810" (Ph.D. diss., Stanford University, 1989), 139–41.

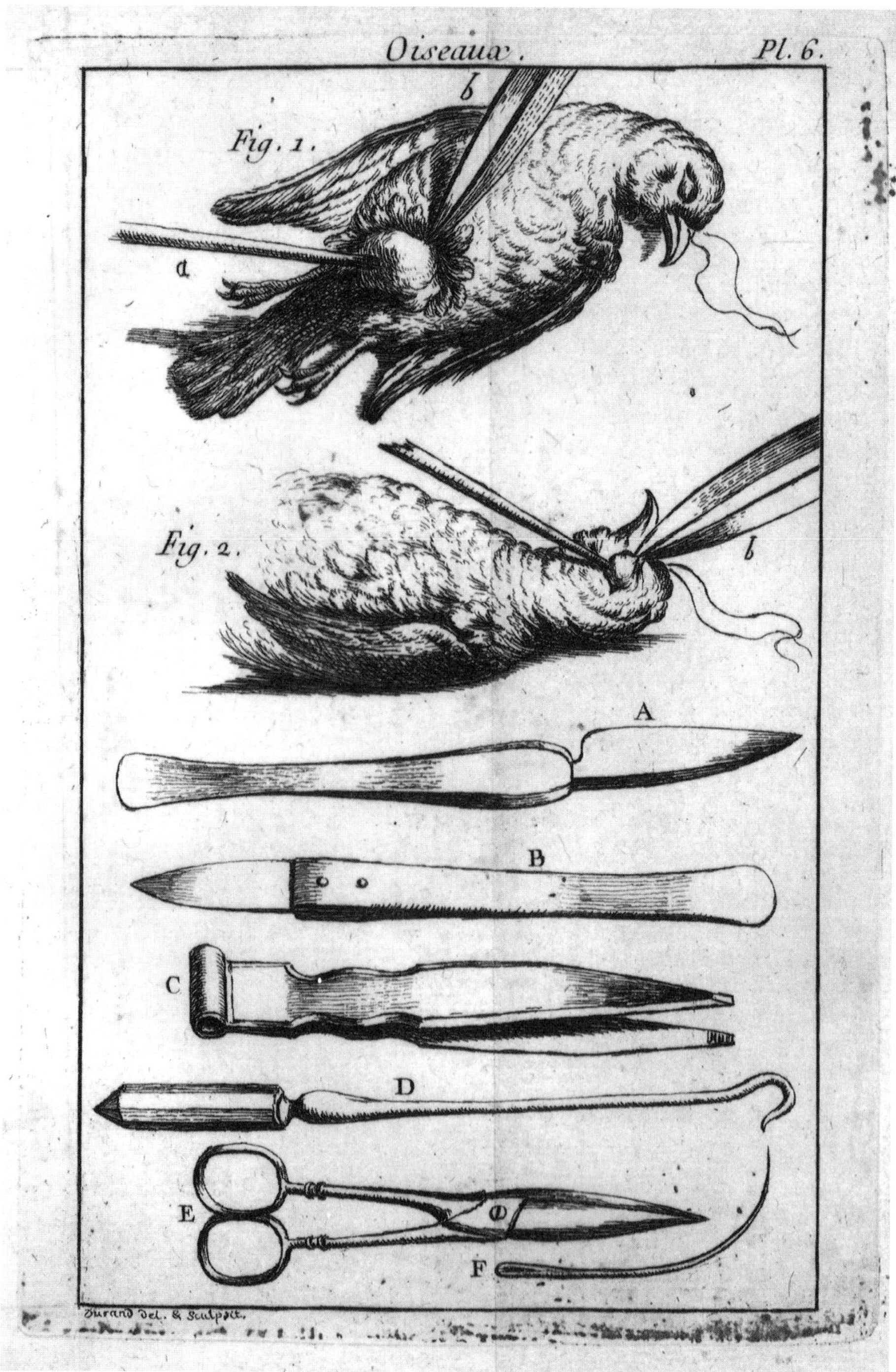

Figure 4. Étienne François Turgot, *Mémoire instructif sur la manière de rassembler, de preparer, de conserver, et d'envoyer les diverses curiosités d'histoire naturelle* (Paris, 1758), plate 6. Huntington Library.

turkey draped over a small chest containing the tools needed to gut and reform the animal (see fig. 2). Above and behind the turkey are cases filled with completed specimens, each of which stands erect. Their heads turn, lift, or bow slightly; their feet are set apart, one foot advanced a step in front of the other. Each specimen stands as if caught in a position held momentarily, no two specimens posed quite alike. If we follow directly up the left side of the canvas, we come to Peale's bald eagle—or white-headed eagle, as he preferred to call it—one of the few specimens from the collection that has survived. The eagle is perched as though waiting, prepared in a moment to spring powerfully into the air and to claim the prey of the fishing hawk. In his public lectures, Peale described the eagle, suggesting that he mounted it in profile to feature its watchful eye:

> It is a mild bird, and unless provoked by hunger never injures other animals. . . . A curious trait in the character of this Eagle is its watching the fishing Hawk, which is seen hovering over the water seeking for food, and after catching a Fish and rising into the air with it, the Eagle, from his superior strength and flight, immediately darts on the Hawk, who in fright lets go his prey, and then the Eagle, with surprising velocity catches the fish before it falls into the water. The strength of the Eagles Eye is really astonishing.[29]

Peale shared this interest in capturing a natural appearance with contemporary naturalists. Mark Catesby, for instance, proclaimed the importance of drawing living birds so that he could seize their characteristic "gestures,"[30] and he used plants (which he also insisted on drawing from life) to enhance such a depiction. In the "Little Sparrow and Purple Bind-Weed," from the first volume of *Natural History of Carolina,* Catesby represented the bird's flexibility and dexterity in gathering insects from the plants around which it lives (fig. 5). Compositionally, the swirling pattern of the bindweed contributes to the sense of motion captured in the bird's twisting, leaning form. By contrast, in the "Yellow Rump" a dead specimen is carefully presented as immobile. Catesby drew the bird suspended from the plant above by a fine strand tied around its foot (fig. 6).[31]

29. Peale, "Vultures, Eagles, and Hawks," lecture 13 of Peale's "Lectures on Natural History" [1799–1800], in Miller, *Collected Papers* (Kraus Microform), IID/12E3–4.
30. Catesby, *Natural History of Carolina,* 1:xi.
31. Ibid., plates 35 and 58.

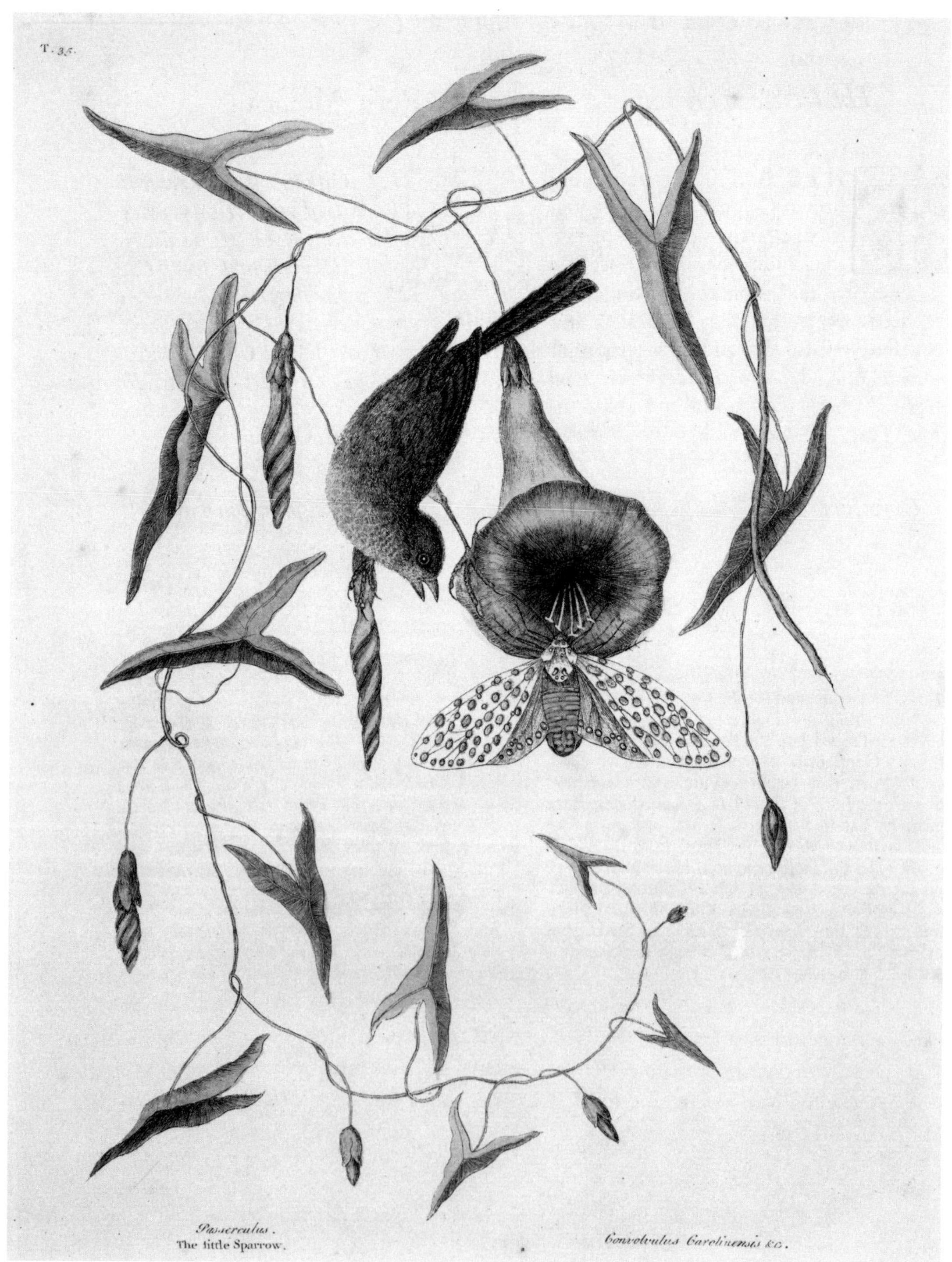

Figure 5. Little sparrow, from Mark Catesby, *The Natural History of Carolina, Florida and the Bahama Islands* (London, 1731–43), volume 1. Huntington Library.

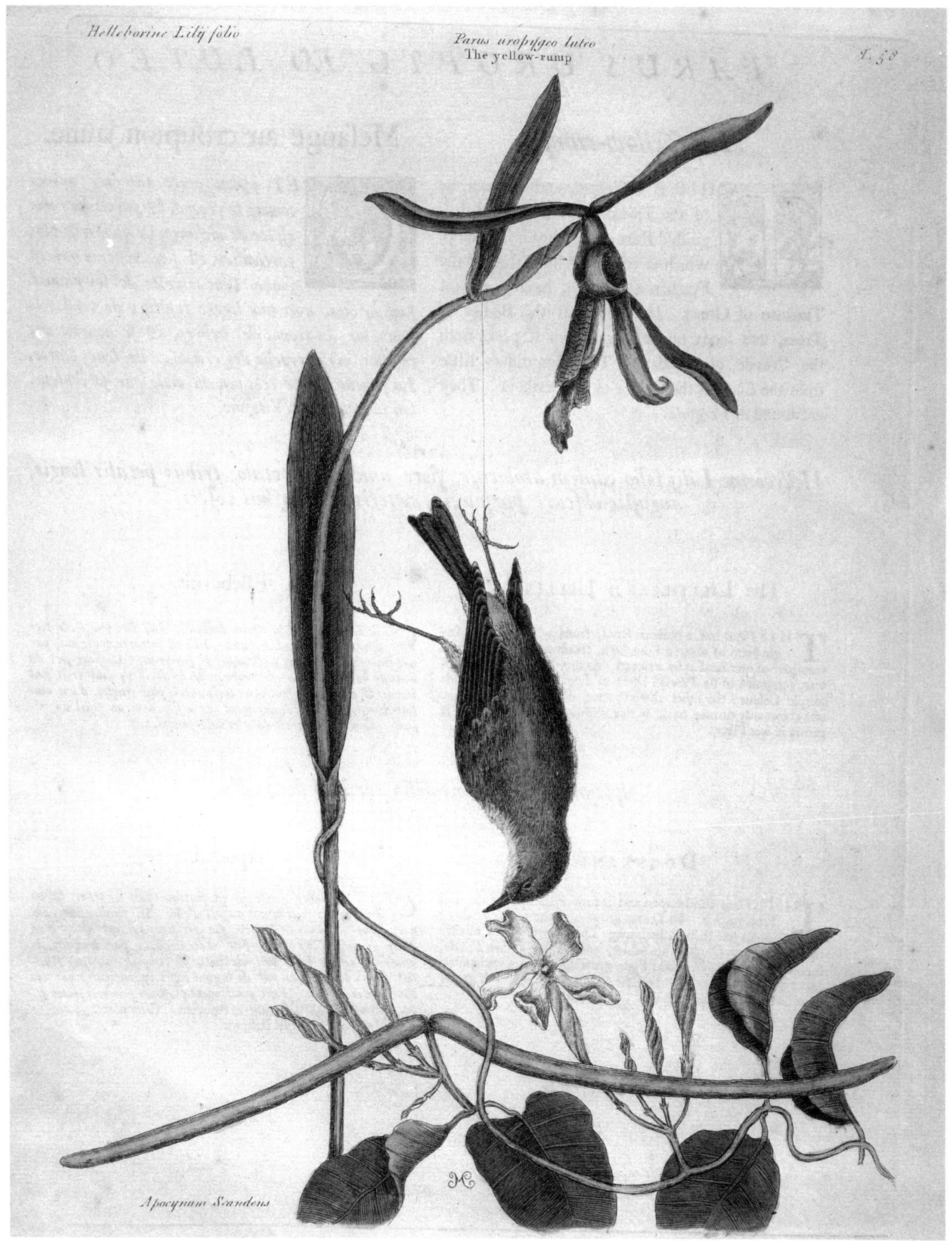

Figure 6. Yellow rump, from *Natural History of Carolina*, volume 1. Huntington Library.

Kuckahn offered elaborate directions and justifications for reconstructing a specimen to look as if it were still alive:

> Without a proper attention to this, however sound your preservation, how vivid soever the plumage may be, your birds are still nothing but meer dead birds; but by a skilful management of *attitudes* and *actions,* you, as it were, animate them, they seem alive, moving and acting. . . . Picking, stretching, feeding, fear, surprize, and fighting, afford peculiar and striking attitudes. . . . Great attention should always be had to the poize of the body: in such positions as a live bird may be supposed to continue some time in, we must take care that the body appears in equilibrium; on the contrary, in fighting and other violent actions, where a forceable motion is to be given, the appearance of equilibrium must be as carefully avoided, for it always conveys the idea of stillness.[32] (Italics added.)

In arguing that specimens be reanimated, many naturalists used words like "character," "attitude," and "gesture" that linked animal displays to human qualities. The taxidermist's art was thus associated with the representation of the human face and figure: the art of the sculptor and portraitist. Turning again to *The Artist in His Museum,* we see that the palette and brushes demonstrate Peale's role as portraitist just as the turkey carcass and tool kit evoke his capacity as taxidermist. Peale's portraits of statesmen and men of letters in the museum stand as parallel expressions to his ornithological character studies. Peale believed that in preserved natural history specimens as well as in portraits of humans, physiognomy conveyed information about the inner character.[33] This belief formed the foundation for building a museum to explicate a natural economy and a social structure fit for a thriving republic.

32. Kuckahn, "Four Letters," 308–11.
33. For another discussion of Peale's use of the palette and brushes, see Stein, "Peale's Expressive Design," 162, 168. On Peale's interest in portraiture and inner character, see Brandon Brame Fortune, "Portraits of Virtue and Genius: Pantheons of Worthies and Public Portraiture in the Early Republic, 1780–1820" (Ph.D. diss., University of North Carolina at Chapel Hill, 1987), 166–222; and David Steinberg, "The Characters of Charles Willson Peale: Portraiture and Social Identity, 1769–1776" (Ph.D. diss., University of Pennsylvania, 1993).

ARRANGEMENT AND DISPLAY

Peale's next step was to create environments that reflected natural habitats. For example, Peale and his assistants painted landscapes to serve as a background for birds displayed in glass cases. In July 1788, for instance, Peale recorded in his diary, "Grinding [water]colours . . . & painting a Landscape in the Bird case with the assistance of my Brother [James] and the latter part of the day untill late placing the Birds in the same."[34] Peale painted such backgrounds throughout the museum's operation, and as late as 1815 he wrote to his son Rubens that he was completing the last landscape for his new installation. Unfortunately, these backgrounds have not survived.

In the early years of the museum's operation, Peale preferred elaborate three-dimensional habitat settings. At the end of his first year in business, Peale wrote to his longtime friend and supporter John Beale Bordley, describing the progress of his efforts: the "ducks & Drakes . . . I have disposed in Various attitudes on . . . Artificial ponds, some Birds & Beasts on trees and some Birds suspended as flying."[35] Contemporary travel accounts attest to the visual impact of the completed display. Historian, minister, and gentleman-naturalist Manasseh Cutler was impressed by the range of Peale's specimens. He described a group of trees on a mound that descended into a pond. Thickets of grass, mineral specimens, and soils offered a variety of settings for the preserved animals. Quadrupeds, birds, reptiles, and fish were arranged on the branches, in the grass, on rocks, and in and around the artificial body of water. The display mingled native and exotic species in a compressed setting that emphasized the variety of animals found in nature and the range of environments that nurture them. For Cutler the display was "romantic and pleasing" and evoked for him an image of Noah's Ark.[36]

Peale gradually reoriented this eclectic exhibition to exemplify two hierarchical schemes, the Linnaean classification of species and the Great Chain of Being. Application of the first helped Peale gain credibility within the scientific community, while the second furthered Peale's interest in presenting a view of the

34. Peale's diary, 11 July 1788 and 24 November 1788, in Miller, *Selected Papers,* 1:513, 547. See also Charles Coleman Sellers, *Charles Willson Peale with Patron and Populace: Transactions of the American Philosophical Society,* 59, pt. 3 (May 1969), 23.
35. Peale to John Beale Bordley, 5 December 1786, in Miller, *Selected Papers,* 1:460; and Peale's diary, 10 August 1788, in Miller, *Selected Papers,* 1:522.
36. Manasseh Cutler, *Life, Journals, and Correspondence of Rev. Manasseh Cutler, LL.D.,* 2 vols., ed. William Parker Cutler and Julia Perkins Cutler (Cincinnati, 1888), 1:261. On Cutler as naturalist, see Manasseh Cutler, "Doctor Cutler's method of preserving the skins of birds," *Collections of the Massachusetts Historical Society* 4 (1795): 9–10.

animal kingdom and society as rank-ordered yet interdependent. This arrangement is evident in *The Artist in His Museum* (see fig. 1). Using Linnaeus as his guide, Peale rearranged the specimens into classes, orders, genera, and species.[37] In the wall of birds, shown on the left side of the painting, for instance, Peale began his display by laying out the species of raptors, including the eagle.

Peale's displays, taken together, demonstrated his belief in the intricate interdependencies among species. All of the basic tenets of the Great Chain of Being—plenitude, continuity, and gradation, in Arthur O. Lovejoy's formulation—were clearly represented. Peale demonstrated the plenitude of nature, its perfection as evident in the variety of forms, in the attempt to represent it exhaustively, and in his perpetual expansion of displays. At the same time, according to the principle of continuity, these abundant forms were ordered as if on a continuum, with each link established by slight variation from the next. The regularity of the twenty-four-by-twenty-inch portraits and the frames that divided the natural history cases contributed formally to the representation of continuity and gradation among species. Each rectangle reserved a space to be occupied by an animal along the continuum. Furthermore, for adherents of the theory, as Lovejoy argued, distinctions of form and purpose explained the position of the links. Peale grouped bats with quadrupeds but placed them over the door leading into the room exhibiting birds.[38] The leap from birds to quadrupeds was in fact gradual, demonstrated by animals that shared the attributes of each.

The idea that hierarchy is universal, natural, and good had great significance for ordering individuals according to gender, race, and socioeconomic rank.[39] In one of Peale's advertisements, the anthropomorphic presentation of the orangutan's pose and visage (fig. 7) suggests its place beside humans in the natural continuum.[40] "How like an old Negro?" Peale asked rhetorically.[41] In other words, the unequal social position of blacks and whites reflected a natural distinction that was necessary to explain the gradations between humans and apes. For Peale, the principles of the Great Chain of Being applied within the human species, not simply across species. The human species existed in abundant variety, each individual occupying a natural position within a social

---

37. On Peale's use of Linnaeus, see Miller, *Selected Papers*, 1:517 n. 121.

38. Arthur O. Lovejoy, *The Great Chain of Being: A Study of the History of an Idea* (Cambridge, Mass., and London, 1936 and 1964), 52 (plenitude), 55–58 (continuity), and 58–59 (gradation); see also Peale, "A Walk," 42.

39. For another argument on the relationship between classification and social order, see Christopher Looby, "The Constitution of Nature: Taxonomy as Politics in Jefferson, Peale, and Bartram," *Early American Literature* 22 (1987): 252–73.

40. Orangutan advertisement, *Claypoole's American Daily Advertiser*, 13 April 1799.

41. Peale, "A Walk," 7; page references given henceforward in the text.

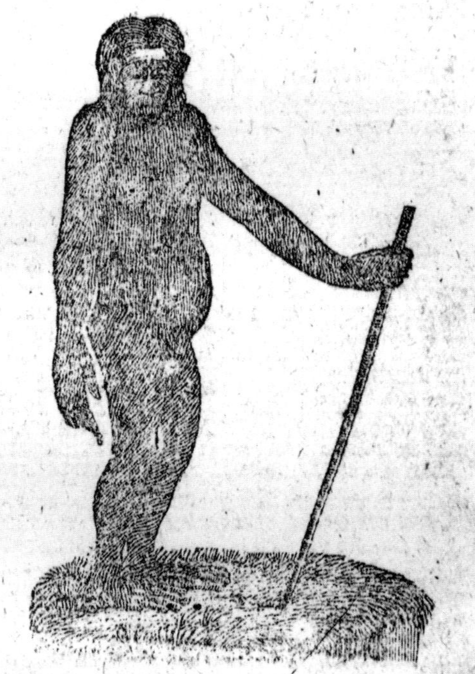

PEALE's MUSEUM.
—
OURANG OUTANG, or WILD MAN
OF THE WOODS.

THIS Curious Animal, so nearly approaching to the human species as to occasion some Philosophers to doubt whether it was not allied to mankind, is now in this useful repository; which is constantly encreasing by the acceſſion of uncommon subjects from all parts of the globe—It consists at present of

QUADRUPEDS—more than 100 of
AMPHIBIOUS animals, upwards of 150 of
BIRDS—near 700.
INSECTS—many thousands.
FISHES—a number of—and a great variety of shells;
MINERALS & FOSSILS—1000 specimens of
PICTURES, and 11 figures of
WAX-WORK,
Representing the persons of various savage nations, all in their proper habits, and surrounded by their implements of war, husbandry, &c.
Also, a powerful ELECTRICAL MACHINE with a medical apparatus. Admittance as usual one quarter of a dollar each time.

Figure 7. Peale's advertisement of the "ourang outang." The Library Company of Philadelphia.

hierarchy. Men and women; refined, middling, and lower sorts; white, black, and Indian—each belonged to a minutely graded, interdependent system.

HARMONY AND HIERARCHY

In catalogues, lectures, newspaper articles, and label text, Peale elucidated the republican implications of this natural system for his audience. Animals were described as useful to humanity as an economic resource, providing food, clothing, and labor; but also and more importantly as a political model. The interdependence of species as well as forms of behavior typical of each species suggested a natural model of harmony and hierarchy upon which to form republican society.[42] That is, animals were both below humans in a system mankind could best manipulate and alongside humans in a system mankind should then interpret for its own edification.

Peale's lectures on natural history were delivered in the winter of 1799–1800 and again the following year. He then condensed the lectures into the form of a manuscript catalogue called "A Walk through the Philadelphia Museum," in which the narrator guides visitors from case to case through the exhibition rooms. The lectures and catalogue disclose yet another type of text Peale used to address the public, for they occasionally quote inscriptions Peale placed on the walls, as do the accounts written by visitors and supporters. Finally, reaching beyond the bounds of the museum, Peale published essays in Philadelphia's newspapers on natural history and on the technological improvements—a more efficient stove, a better bridge—shown in his displays.

The abundance of nature predicted the future wealth of the American republic, and the museum taught ways to convert those resources. Foxes, ermine, bears, and muskrats all provided valuable furs (see "A Walk," pp. 19, 23, 24, and 30). Cloth could be woven from the hair of the llama (p. 35) and the buffalo (p. 40). Ducks, pheasants, and turkeys were identified as particularly delicious game birds, and bear and buffalo meat were also pronounced appetizing (pp. 82, 24, 40).[43] Peale favored the domestication of many species. The ox earned high praise as the "support of husbandry and the strength of Agriculture" (pp. 39–40). The buffalo, larger and faster than the ox, promised

42. For an exploration of Peale's view that harmony was socially ameliorative, see Lillian B. Miller, "Charles Willson Peale: A Life of Harmony and Purpose," in Edgar P. Richardson et al., eds., *Charles Willson Peale and His World* (New York, 1983), 170–233.

43. See "Lectures on Natural History" [1799–1800], in Miller, *Collected Papers* (Kraus Microform), lecture 28, p. 7, IID/20; and lecture 29, typescript, IID/21.

to be a superior draft animal (p. 40). Other animals helped the American farmer without having to be trained. Owls depleted poultry stocks, but they also killed rats (p. 46). Grackles fed on many insects, worms, and reptiles that destroyed crops; the corn they also consumed was a small price to pay for eliminating these other pests (p. 57).

When Peale described the behavior of animals as suggestive of ideal human social organization, he attributed human traits, usually positive ones, to dozens of species. While Peale recognized negative characteristics—vanity, obstinacy, and indiscretion—among some animals, he far more often identified their redeeming qualities (pp. 9, 35, 82). Peale understood nature to embody an essentially cooperative rather than a competitive existence. Relationships within and across species exhibited for Peale a natural harmony, one to be emulated by American society both domestically and publicly. A brief list of praiseworthy traits and the animals that possess them suggests Peale's hopes for promoting accord in the home. The mongoose and coatimondi were notably playful, and the macaw was good-humored (pp. 21, 49). The dog was, of course, faithful (p. 18). The Indian musk and the entire genus of corvus, consisting of crows and jays, were recognized as sociable, while monkeys, cockatoos, and swans were described as affectionate (pp. 36, 53, 10, 49, 81). Some species were admired for their motherly attention to their offspring, including apes, sloths, seals, oppossums, and pelicans (pp. 9, 12, 17, 25, 90). The nuthatch was singled out among birds as exemplary of marital affection (pp. 70–71). Adding a nationalist dimension to these anthropomorphic observations, Peale noted the faithfulness of the American cuckoo in contrast to the European species. The Old World female cuckoo gained a reputation for infidelity, perpetuated by the word *cuckold,* by laying her eggs in another bird's nest. By contrast, Peale claimed, "the American Cuckow . . . build their own nest . . . [and] foster their young;—they diligently chant their soft notes to sweeten the care of incubation and we are proud to believe that they are faithful and constant to each other" (p. 60).

Peale's vision of social harmony focused upon animals that were, he claimed, gentle by nature.[44] Peale singled out the talapoin among monkeys for its "extremely gentle manners." From such behavior by individuals followed enormous social rewards; Peale claimed "that it not only gives peace and comfort to the possessor, but distributes happiness all around; It disarms the passionate; puts an end to strife and its baneful consequences and sometimes arrests the

---

44. Richard Bushman demonstrates the importance of manners in defining American gentility from the early colonial period to the end of nineteenth century. Bushman's analysis suggests that manners simultaneously promoted ease of address and social hierarchy; see *The Refinement of America: Persons, Houses, Cities* (New York, 1993).

direful purposes of Madmen."[45] Animals that feed on grass, especially ruminating animals, exhibited a natural capacity toward a gentle disposition. But even
carnivorous animals were made violent only by hunger, Peale proposed. Despite
their reputation for ferocity, the cougar, the wolf, and the fox were grouped
together as gentle animals ("A Walk," p. 19).[46] The bear was the exception that
proved the rule, and Peale characterized it as not only "savage" and but also
"solitary," suggesting that its violence was a function of its distance from civilization and society.[47] More than a vision of social harmony, this also would
seem to be a republican formulation of power: while different species possess
greater or lesser ability to overpower one another, Peale admired the restraint of
animals that do not wield force wantonly.

Clearly, predatory animals posed the greatest challenge to Peale's natural
model of social harmony, but his treatment of the eagle suggests a resolution
of this problem. His representation of the eagle stands in marked contrast to
that of other naturalists. Peale's mount, as depicted in *The Artist in His Museum*,
suggests a system of interdependent species (see fig. 2); whereas Audubon's
watercolor (frontispiece), for example, depicts a violent struggle between two
species. Recall that the implied narrative of Peale's exhibit is that the eagle is
waiting for a hawk to catch a fish, after which the two birds will compete for
the prey.[48] But these are not the only layers in the system. The king bird challenges both eagles and hawks, Peale writes, and "thus he may be considered the
protector of our Poultry" ("A Walk," p. 48). In Audubon's understanding of natural economy, the outcome seems inevitable, whereas Peale proposes a check
against the power of the great predators. In another exhibit featuring an eagle,
Peale took a step further in asserting that powerful animals need not destroy
weaker ones. In 1797 he advertised "a singular association" between a rooster
and an eagle (although the museum was best known for its preserved specimens, Peale also maintained a menagerie of living animals). The chicken,
"since the death of its own species, has deserted its wonted cage, and regularly
roosts, each night, with one of the Eagles;—This fact is astonishing—as, the
Eagle is well known to be the greatest enemy to poultry."[49]

45. Peale, "Orang-Outang, Baboons and Monkies," in Miller, *Collected Papers* (Kraus Microform),
    lecture 1, p. 22, IID/5.
46. On the wolf, see also Peale, "Ant Eater, Manis, Armadilla, Sea Lion, Sea Bear, Seal, Hyena, Jackal &
    Wolf," ibid., lecture 3, pp. 26–27, IID/6.
47. Peale, "Coati-Mondi, Otter, Pole-Cat, Minx, Stoat, Weasel, Ferret, Ermine, Bear, Raccoon, and
    Opossum," ibid., lecture 5, typescript, p. 12, IID/7.
48. Peale, "Vultures, Eagles, and Hawks," ibid., lecture 13, p. 5, IID/12E3–4.
49. "A Singular Association," *Claypoole's American Daily Advertiser,* 6 January 1797.

In his writing as in his museum, Peale made the case that every animal was ideally suited to its position within the continuum of the Great Chain of Being. Peale repeatedly set himself in opposition to the French naturalist Georges-Louis Leclerc, Comte de Buffon, who rejected classification as arbitrary and argued that nature was characterized instead by disorder. The guinea pig was gentle, according to Buffon, but capable only of reproducing—not of accomplishing good. Peale chastised Buffon for this view, writing, "I do not admire this degradation of any species. . . . It may be witty. I am sure it is not pious" ("A Walk," p. 29). Even members of the lowliest species served a valuable purpose, and therefore their perpetuation was necessary. Buffon described the existence of the woodpecker as irredeemably painful and slavelike, but Peale countered that these archetypal workers were uniquely formed to accomplish their life-sustaining tasks (pp. 62–66). Moreover, employing the portraitist's assumption of the relationship between physiognomy and character, Peale argued that the woodpecker's face "is rather dainty, than coarse." Extracting the moral from this case, Peale continued, "The goodness of providence is manifested to all his creatures; each enjoy life in its fullest extent:— None are made to be miserable, or even unhappy.—The sphere of each is measured with equal goodness. and it is those only, who suffer, that step aside from Natures paths."[50] In human terms, Peale's perceptions of hierarchy attributed social distinctions to natural, even providential, order. While this precluded social mobility, it commanded the lofty to respect and appreciate their inferiors. To do otherwise would be unnatural and against the will of God.[51]

"ASK THE BEASTS, AND THEY SHALL TEACH THEE"

On the walls of the museum, Peale inscribed quotations to inspire Christian contemplation of divine works in nature. His lectures and advertisements quoted directly from biblical passages, and he actively sought the support of Philadelphia's clergymen in promoting the museum. Many of them endorsed the museum's teachings, and they consistently argued that there was no contradiction between scientific knowledge and Christian piety.[52] Among those

---

50. Peale wrote, for example, "The happiness of many Individuals is attached to the happiness of mankind, and he is under obligations to exert himself for the general good, because his own depends on it"; see "Orang-Outang, Baboons and Monkies," Miller, *Collected Papers* (Kraus Microform), lecture 1, p. 7, IID/4.
51. See *Dictionary of Scientific Biography*, 2:576–82.
52. Supportive ministers included Episcopalians William White, Samuel Magaw, John Andrews, Joseph Pilmore, Henry L. Waddell, and James Abercrombie; Presbyterians John Ewing, John Blair Smith, and Ashbel Green; Associate Church minister William Marshall; Baptist Thomas Fleeson; and Lutheran Henry Helmuth (see Brigham, "'A World in Miniature,'" 452).

ministers, Swedish Lutheran leader Nicholas Collin was the most active pro-
ponent of the religious significance of Peale's natural history displays.

Beginning in 1788, Peale distributed tickets depicting a banner that read,
"The Birds & Beasts will teach thee!," a paraphrase from Job 12:7 (fig. 8). This
passage thus greeted every visitor, and it continued to be important in promoting the museum. Collin featured it in a series of essays he wrote in support
of Peale, published in December 1800 to encourage attendance at Peale's lectures on natural history. In his first essay, Collin argued that the study of nature
was important chiefly because it facilitated the "direct promotion of Religion."
Collin supported his belief in the consistency between science and Christianity
with a number of biblical quotations, including the one from Job, expanded:

> Ask the beasts, and they shall teach thee; and the fowls of the
> air, and they shall tell thee; and the fishes of the sea shall
> declare unto thee. Who knoweth not in all these, that the
> hand of the Lord hath wrought this? In whose hand is the soul
> of every living thing, and the breath of all mankind?[53]

The lessons to be learned through attendance at the museum were complementary to those gained through other aspects of a pious life.

In the principle of plenitude, Peale found evidence not only of the Great
Chain of Being but also of the work of God. In his opening lecture Peale
declared the variety and fullness of nature to "display the wisdom and Goodness
of a Benificent God." He quoted Psalm 104, "O! Lord how manifold are thy
works; in wisdom hast thou made them all; the Earth is full of thy riches.[54] The
wealth of the new republic was based on resources to be exploited, but it also
embodied divine providence. Peale recalled for his listeners God's charge to
Noah after the flood:

> And the fear of you, and the dread of you shall be upon every
> Beast of the Earth, and upon every fowl of the air, upon all
> that moveth upon the Earth, and upon all the fishes of the
> Sea: into your hands are they delivered.[55]

53. Nicholas Collin, "Remarks on the utility of Mr. Peale's proposed Lectures in the Museum," 6 parts,
      Poulson's American Daily Advertiser, 17–24 December 1800.
54. Peale, "Orang-Outang, Baboons and Monkies," Miller, *Collected Papers* (Kraus Microform), lecture 1,
      p. 2, IID/4. I am grateful to James Thorpe for help in identifying the source of this passage.
55. Genesis 9:2, quoted in Peale, ibid., lecture 1, pp. 6–7, IID/4.

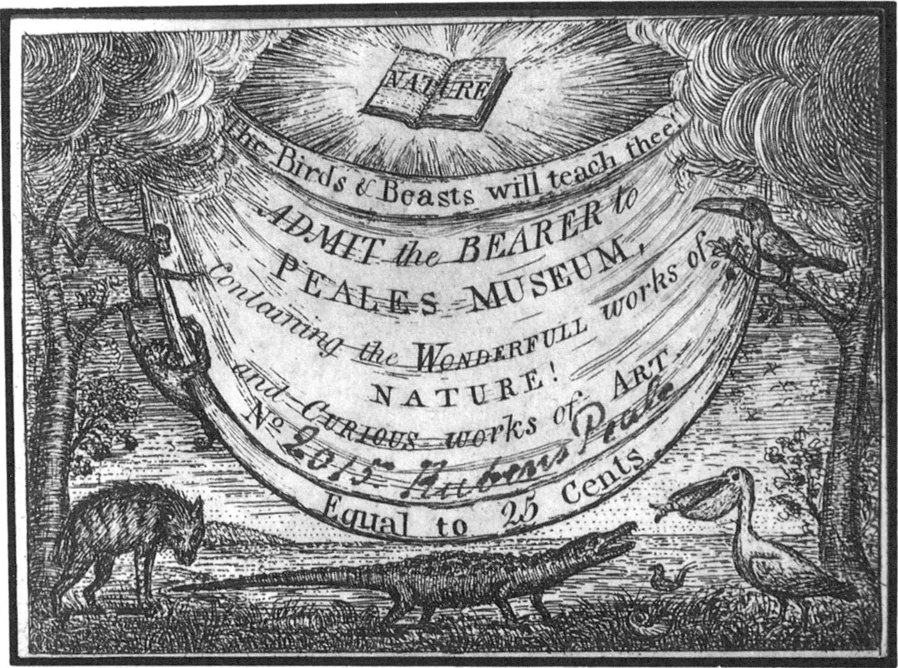

Figure 8. Entrance ticket to the Peale Museum (1788). Private collection; photo courtesy of Hirschl & Adler Galleries, New York.

However, this grant of control carried with it the responsibility of stewardship. Despite the limits of lesser animals, God provided for the care and sustenance of all. Collin, quoting Matthew, suggested this aspect of God's plan to Peale's audience: "Behold the fowls of the air: for they sow not, neither do they reap, nor gather into barns; yet our Heavenly Father feedeth them."[56] The republican implication of this passage may have been to remind the better sort of their duty to sustain and care for the lower sorts.

God's intention was also evident in the fittedness of all creatures to their place in the natural hierarchy. Despite the awkward appearance of the toucan, for instance, Peale argued that "on a closer view we find the perfection of the work of Creation; in providing suitable organs for the support and also happiness of each individual being: each species fitted to fill their various stations allotted them in the great scale of Nature, and the more we inspect into her ways, the more we shall have abundant cause to lift up our hearts and minds in love and admiration of the great first cause!" ("A Walk," p. 50). Reference

56. Collin, "Remarks," pt. 1, *Poulson's American Daily Advertiser,* 17 December 1800.

to the Bible added the weight of God's will to Peale's vision of a republic founded upon natural principles.

In constructing his natural history displays, Peale drew upon all of his skills as an artist, a term that I use here in the broadest sense to mean the human transformation of materials. He hunted animals, crafted specimens into lifelike forms, and created habitats in which to exhibit them. Through his writings, Peale elaborated on his intentions in "arranging the stuff of nature." The goal of his interpretive program was to present nature as inherently harmonious and hierarchical. Peale encouraged citizens to model their economic, social, and political life after these two natural principles, and he and his colleagues urged that such order was also God's will. I conclude by quoting a passage from Reverend Collin's essays on Peale, in which he extolled the museum's success in presenting harmony and hierarchy as naturally and divinely ordained social values:

> Knowledge and goodness are by Him distributed in a great variety of degrees from the upper to the lowest ranks! discernible in their domestic plans, and social systems. Good will to the species—reciprocal offices among fellow citizens—parental love—filial affection, conjugal tenderness—are general and in the degrees most suitable to the condition of each kind.
>
> . . . A Museum stored with specimens of quadrupeds, birds, fishes, amphibia, insects, animals, &c. from all parts of the terraqueous globe, is a miniature of it; and a temple which no thinking person can frequent without adoration of the Creator.[57]

Worcester Art Museum

---

57. Collin, "Remarks," pt. 2, *Poulson's American Daily Advertiser,* 18 December 1800.

"Your garden must be a museum to you": Early American Botanic Gardens

THERESE O'MALLEY

Often coupled with a museum or natural history collection, the botanic garden was promoted as an essential "national object" by such figures as George Washington, Thomas Jefferson, John Quincy Adams, and Charles Willson Peale.[1] In this essay I consider how the botanic garden achieved this status and why these cultural and political leaders found it useful for their purposes. First I describe some of the earliest efforts to found botanic gardens in America—how they were established and how they developed in response to changing scientific and cultural concerns. In the second part of the essay I consider a succession of botanic gardens planned for the National Mall in Washington, D.C. I shall argue that the botanic garden became the quintessential expression of both garden art and scientific inquiry and, moreover, that it was particularly useful during the formative decades of American nationality, when the ruling elite was trying to establish superiority over Europe and to create a national identity celebrating the natural richness of the new continent as well as achievements in the sciences and the arts.

Charles Willson Peale—artist, scientist, and inventor—in many ways typified late-Enlightenment intellectual and cultural pursuits.[2] In his natural history museum he attempted to demonstrate the interrelatedness of natural phenomena in what he called his "world in Miniature" (see fig. 1, *The Artist in His Museum,* in David Brigham's essay in this volume). He juxtaposed animals, bones, minerals, and ethnographic material in displays organized according to a hierarchical order, with portraits of heroes of the Revolution presiding over the "lower classes"—predators, birds, fish, and minerals. Peale's museum made visible his perception that there was an order of things encompassing not only the

1. Richard Rathburn, "The Columbian Institution for the Promotion of Arts and Sciences," *United States National Museum Bulletin* 101 (1971): 37–39.
2. Sidney Hart and David C. Ward, "The Waning of an Enlightenment Ideal," in Lillian Miller and David C. Ward, eds., *New Perspectives on Charles Willson Peale* (Pittsburgh, 1991), 219–36.

≈ 35

natural world but also social structure. Indeed, until 1800, when the city of Washington was made the capital, Peale's museum shared Independence Hall in Philadelphia with the newly formed federal government of the United States. Beyond its scientific and historical significance, the museum was a powerful statement of organization within the new nation: it was the first institution in the new age of democracy to combine scientific system and broad educational intent.[3]

What Peale achieved for ornithology, biology, geology, and paleontology in his museum could be done for botany only in a garden. Peale was frustrated in his attempt to display botanic material in his museum because plants did not retain a natural appearance once preserved. Only a garden where living plants were studied could serve the needs of the new empirical sciences. Attempts to use the Pennsylvania State House Yard for this purpose were unsuccessful, so it was not until late in his life, when he purchased his own property, Belfield (fig. 9), that he was able to create a botanic display. Peale's advice to his colleague Thomas Jefferson, "your garden must be a museum to you," suggests that at Belfield, Peale was trying to create a living museum of plants.[4] We can understand Peale's improvement of his farm as an attempt to perfect and to order the natural world, just as he did by means of the natural history objects in his museum. Belfield was a place in which the didactic, the scientific, and the aesthetic were brought together, embodying Peale's life's work.

The proprietors of other large private gardens in Philadelphia, such as The Woodlands and Lemon Hill, established some of the great early botanic collections. It was reported that at The Woodlands, for example, there "was not a rare plant in Europe, Asia, Africa, from China and from the islands in the south Seas" which William Hamilton had not procured.[5] Such collections were built by members of an international network of scientists and gardeners who carried on an extensive exchange of plant material. These estates represented not only the peak of botanic collecting, achieved during and just following the Revolutionary War years, but also the fruitful patronage of scientists by wealthy landowners.

3.   See Gary Kulik, "Designing the Past: History-Museum Exhibitions from Peale to the Present," in Warren Leon and Roy Rosenzweig eds., *History Museums in the United States: A Critical Assessment* (Urbana, Ill., 1989), 6.

4.   Peale to Thomas Jefferson, 22 March 1812; see Charles Coleman Sellers, *Charles Willson Peale* (Philadelphia, 1969); and O'Malley, "Belfield in American Garden History," in Miller and Ward, *New Perspectives on Peale*, 276.

5.   E. McLean, "Town and Country Gardens in Eighteenth-Century Philadelphia" *Eighteenth-Century Life* 8 (1983): 145. Thomas Jefferson entrusted to Hamilton the living material brought back from the Lewis and Clark expeditions.

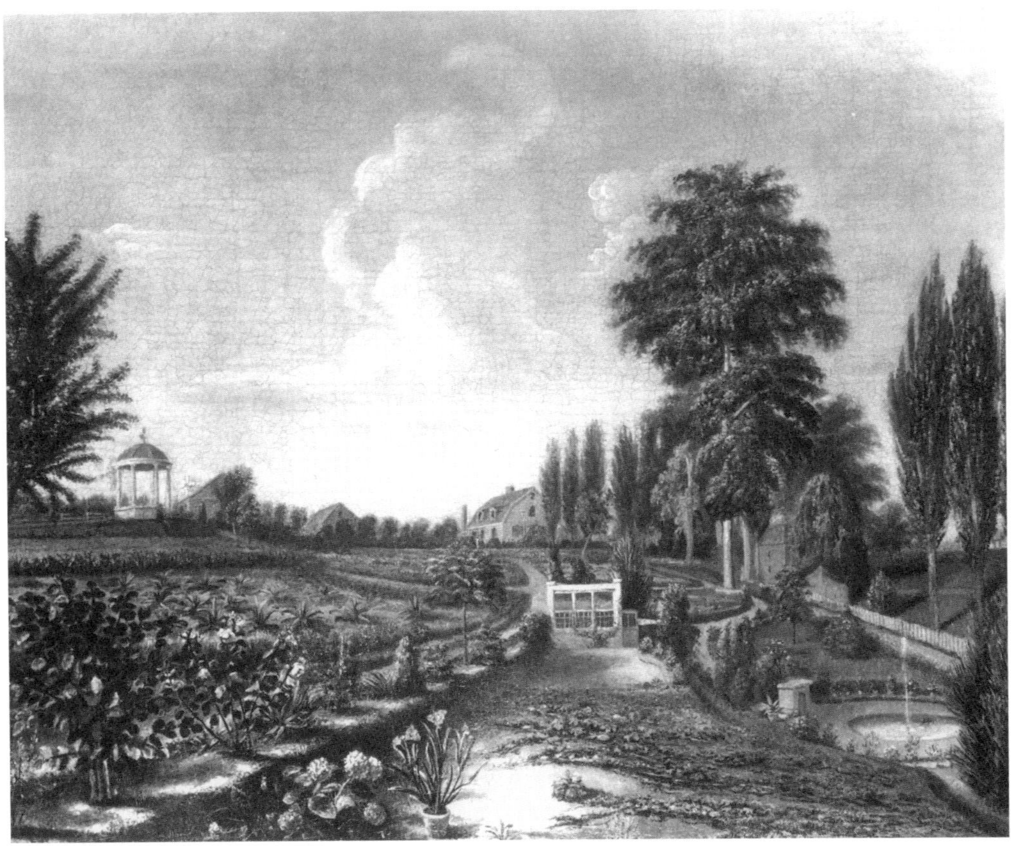

Figure 9. Charles Willson Peale, *View of the Garden at Belfield* (1816); oil on canvas, 29.9 x 40.6 cm. Private collection.

The history of botanic gardens in America over the next century reflects a confluence of changes in the perception of the natural world and in landscape aesthetics.[6] Developments in the ecological sciences influenced the organization of botanic gardens and consequently their design. The history of gardens provides a vivid example of the shift from a static view of what has been called the Great Chain of Being and the fixity of species to a more inclusive pre-evolutionary perception of the complexity of the natural world. Greater knowledge about the natural habitats of plants, and how to recreate them, as well as the aesthetic desire to see nature unrestrained by artificial devices, converged to influence garden design in general and botanic gardens and natural history collections specifically.

6. See Therese O'Malley, "Art and Science in the Design of Botanic Gardens, 1730–1830," in John Dixon Hunt, ed., *Garden History: Issues, Approaches, Methods* (Washington, D.C., 1992), 279–302.

The botanic garden developed in part as a response to the new empirical sciences and to the new understanding they fostered of the influence of environmental conditions. But this was applied to human as well as to plant life, and the notion that environment exerted an influence on human character had a long tradition in social theory. Hector St. Jean de Crevecoeur, whose writings exemplify eighteenth-century formulations of the idea, wrote in reference to the American farmer:

> Men are like plants, the goodness and flavor of the fruit proceeds from the peculiar soil and exposition in which they grow. We are nothing but what we derive from the air we breathe, the climate we inhabit, the government we obey, the system of religion we profess, and the nature of our employment.[7]

This quotation is a particularly cogent expression of environmental determinism, pervasive in America, where a new nation that proclaimed itself to be in compliance with "natural law" was in its formative stages. The botanic garden was perhaps the clearest attempt to construct an improving environment.

In the same period, gardens were acquiring greater educational and economic significance. The gardens at Harvard (fig. 10), Yale, Princeton, Columbia (fig. 11), the Medical College of South Carolina, and the University of Pennsylvania were established within the first two decades of the nineteenth century. In 1811, Benjamin Waterhouse's *The Botanist* proclaimed a new value placed on empirical research and on education, and the simultaneously aesthetic and scientific capacity of the botanic garden to advance research and education. Waterhouse, professor of medicine at Harvard from 1783 to 1812, exhorted the student of botany to pass from the closet to the gardens and fields in order to study the scriptures of nature rather than books written by men. "The utility of these institutions is self-evident," he wrote:

> By public gardens, medicinal plants are at the command of the teacher in every lesson, the eye and the mind are perpetually gratified with the succession of curious, scarce and exotic luxuries, here the botanist can compare the doubtful species, and examine them through all the stages of growth, with those to which they are allied.[8]

---

7.   J. Hector St. Jean de Crevecoeur, *Letters from an American Farmer,* originally published 1792; reprint, ed. Albert Stone (New York, 1981), 71.

8.   Benjamin Waterhouse, article 11 in *The Botanist* (Boston, 1811), 109.

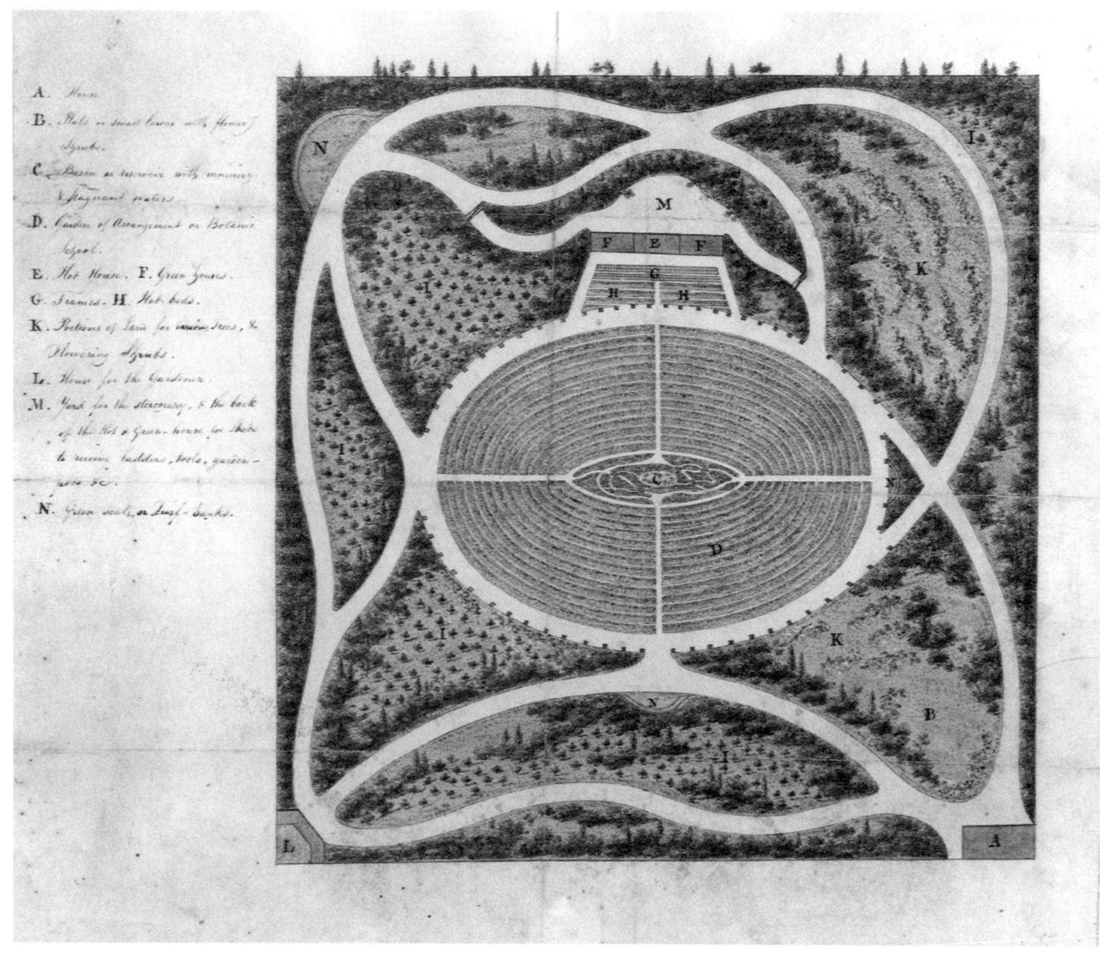

Figure 10. Early plan of the Harvard Botanic Garden. Library of the Gray Herbarium, the Botany Libraries, Harvard University.

Waterhouse referred to botanic gardens as "a new species of luxury to the botanist," but he pointedly described them as public—meaning that they were to be for the edification of every citizen. The distinguishing characteristic of botanic studies, repeatedly emphasized by American scientists, educators, and public figures, was accessibility: it was by means of gardens that citizens, whether educated or not, could acquire knowledge of plants. Botanic gardens, hitherto primarily medicinal in purpose, became in the early nineteenth century museums of living plants that were centers for research and education, display and delight.

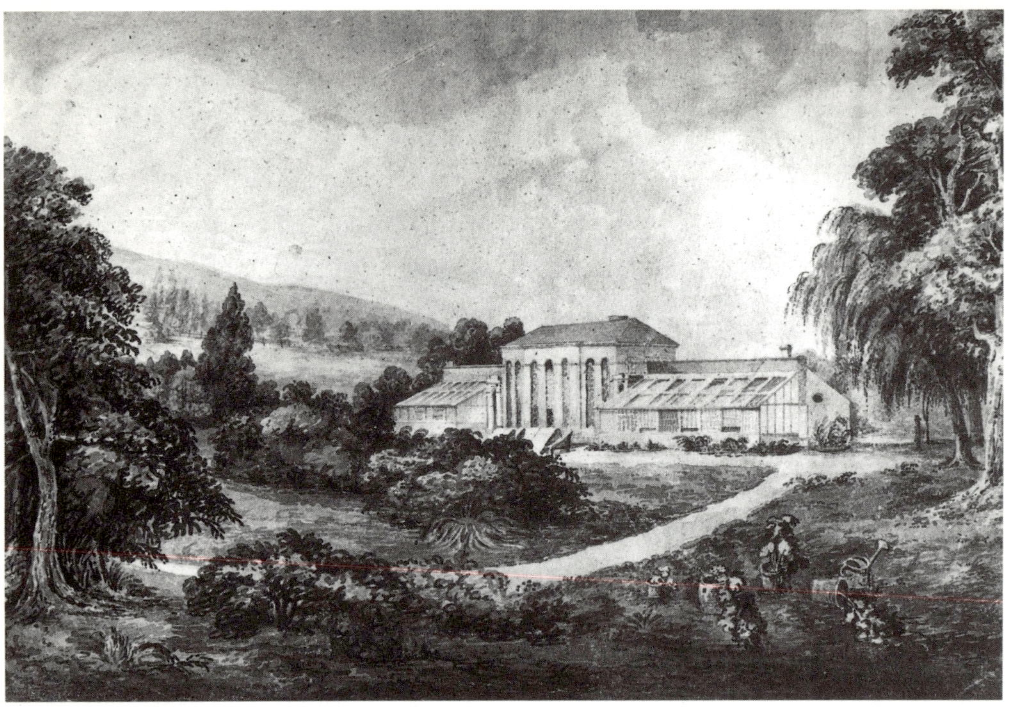

Figure 11. Hugh Reinagle, *View of Elgin Garden on Fifth Avenue* (ca. 1812), sepia drawing. I. N. Phelps Stokes Collection, Miriam and Ira D. Wallach Division of Art, Prints, and Photographs, New York Public Library, Astor, Lenox, and Tilden Foundations.

In early American botanic gardens—especially those found in public spaces—scientific, agricultural, and ornamental interests were inextricably linked. But the naturalization and cultivation of useful plants taking place in these gardens were recognized in particular as essential to the national economy. These gardens were showcases of the natural raw material that was integral to the commercial success of the larger community. Agriculture, always foremost in the American economy, was no longer seen to be in opposition to manufacturing and industrial pursuits, as it had been in the previous century. Now both agriculture and industry could work together to bring the materials of nature into ultimate production. This utilitarian conception of nature appealed to this country's leaders, who were anxious to teach both self-sufficiency and superiority in terms of the international marketplace. Seen in these terms, industrialized society would produce no conflict between agrarian interests and ambitious factory output. It was all part of the same economic model.

George Washington and Thomas Jefferson, as well as a few other members of the landed elite, expressed their concern about the backward agricultural methods practiced by the majority of farmers in this country at the end of the eighteenth century. They pointed to the neglect of crop rotation, the absence of labor-saving tools, and the poor condition of livestock and land as severe obstacles to American competitiveness. The need to promote research and education in so-called scientific agriculture—as well as to foster the taste for rural aesthetics—led to the organization of many agricultural societies immediately following the Revolution. By 1800, such societies had been established by wealthy landowners and political and intellectual leaders in most of the major cities: Philadelphia, New Haven, Charleston, South Carolina, Boston, and New York.[9] The interest in creating botanic gardens in the city of Washington in part grew out of this desire to improve the practice of agriculture.

Proposals for a botanic garden on the National Mall date from the very inception of the utopian and visionary plan for the new capital. Plans for a national botanic garden on the Mall were first discussed in the 1790s, in correspondence between George Washington and his commissioners, and between Thomas Jefferson and his colleagues in artistic and scientific circles (figs. 12, 13).[10] In their discussion of public grounds in the city, the commissioners agreed that a botanic garden—in addition to a national university and mint—was essential.[11] For Washington and Jefferson, a botanic garden was a necessary part of the agrarian utopia that they hoped to found. Expressed in that ideal, democratic in essence, was the obligation to provide education of all kinds—social, moral, and intellectual. Botanic gardens could best realize that objective because they were a "perfected environment" constructed according to scientific system, yet accessible to all. The belief that social and moral benefits were conferred by this garden type contributed to the desire to create not only a botanically ornamented seat of government but also a model didactic environment where one would be taught the principles of botany and landscape art.

Over the first decade of the capital city's existence, many proposals to create botanic gardens on the Mall were put forward. In one such scheme published in the *Washington Expositor*, the writer claimed that although the public grounds could be put to strictly functional use by simple cultivation, the presence of

---

9. Richard Beale Davis, *Intellectual Life in Jefferson's Virginia, 1790–1830* (Knoxville, 1972), 151–53. See also Harold T. Pinckett, "Early Agricultural Societies in the District of Columbia," *Records of the Columbia Historical Society* 51–52 (1951–52): 32.
10. Rathburn, "The Columbian Institution," 37–39.
11. W. B. Bryan, *A History of the National Capitol*, vol. 1 (New York, 1914–17), 276.

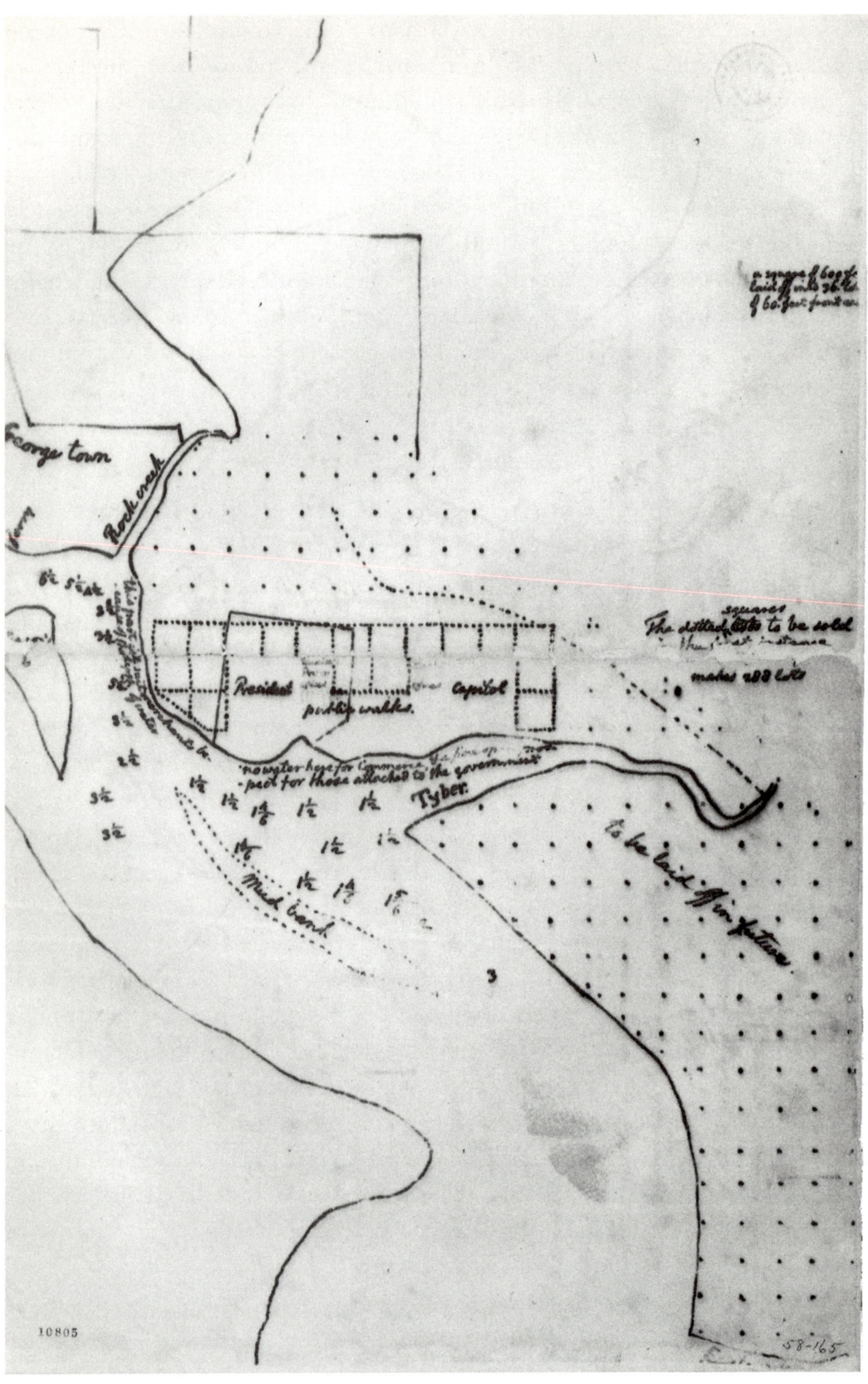

Figure 12. Thomas Jefferson, Sketch of plan for Washington (March 1791). Library of Congress.

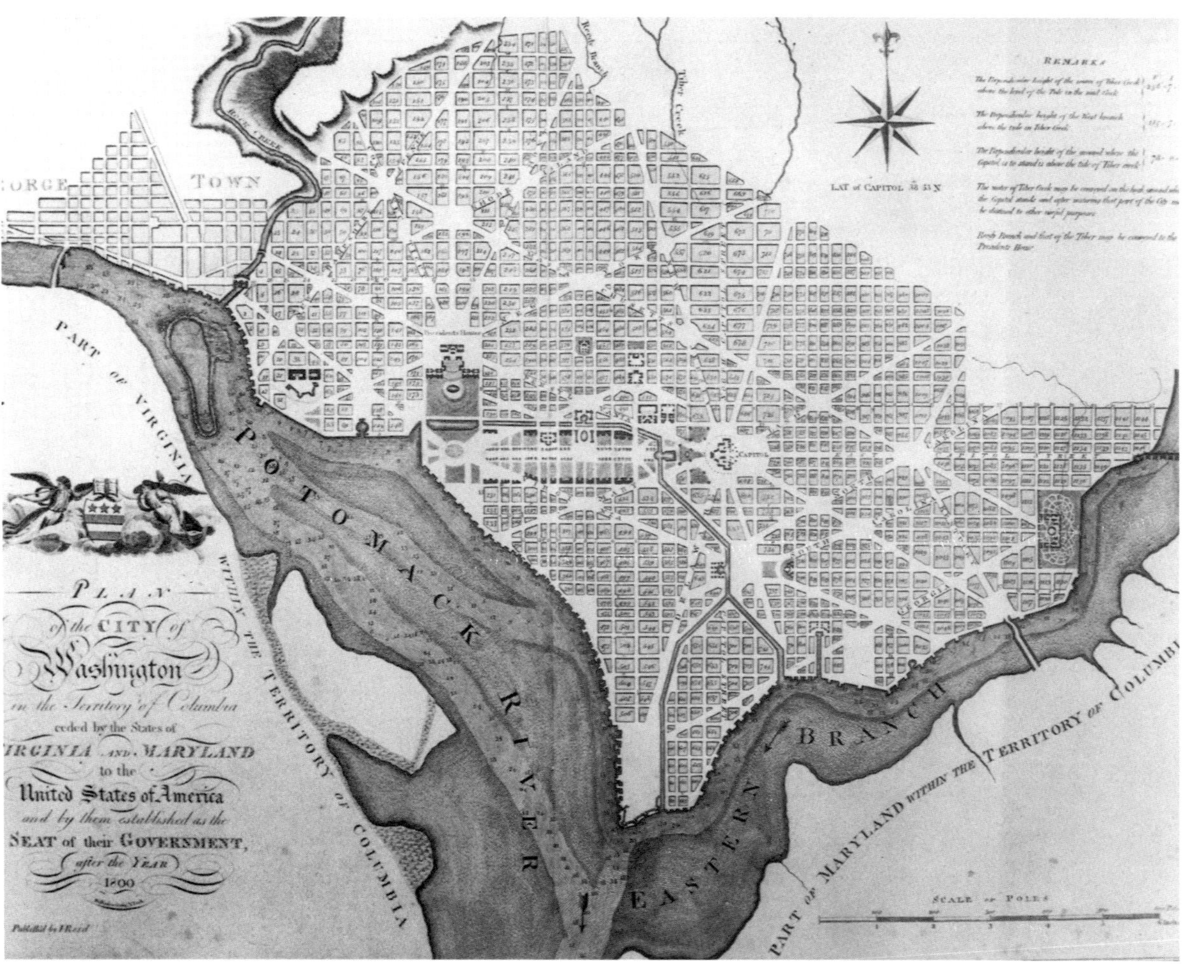

Figure 13. Plan of the city of Washington, engraved by William Rollinson (1795). White House Collection.

botanic gardens might also serve an aesthetic purpose.[12] Other proposals focused on the botanic garden's cultural and educational possibilities, however, and its capacity to demonstrate American excellence. In one proposal for a botanic garden published in a Washington newspaper in 1801, the writer emphasized education over ornamentation, claiming that even the illiterate man would be capable of instruction through the direct observation of plants. He further urged that a garden of indigenous material, so far superior to one based on Old World varieties, could exert a profoundly moral influence upon citizens.[13] A letter to

12. *Washington Expositor,* no. 1 (2 January 1808).
13. Quoted in Rathburn, "The Columbian Institute," 37.

the editor of the *National Magazine* insisted on the necessity of a botanic garden to promote "scientific knowledge of the various branches of agriculture." Further, the writer suggested, in a garden of fifty acres the collection of native specimens would be not only scientifically comprehensive but also useful to American botanists in establishing economic and cultural independence from Europe. While native specimens were to be principally celebrated, foreign specimens would also be exhibited as proof of the flexibility of the American environment and the skill of American gardeners in naturalizing them.[14] A proposal of 1808 made the nationalistic argument once again, stating, "To render us independent we must raise and naturalize those plants, the products of which are by custom, rendered necessary to our comfort and convenience." This writer also expressed the common patriotic notion that the huge continent could support any plant better than a foreign nation could. He asserted with confidence the connection between gardens and government, and the role of the botanic garden as a "national object":

> Gardens and nurseries, where the chemist, botanist, and agriculturalist can have free access at all seasons will it is hoped, now become of peculiar interest to the patriot and legislator. . . . For objects of this nature there is certainly no place better adapted than the seat of the general government. . . . Within the limits of the federal seat there are large and ample reservations for public gardens and other national objects, which may advantageously be applied to the purposes of a botanical garden, a public nursery and an agricultural farm.[15]

Thus in proposals for the National Mall, botanic gardens were deemed essential for the purposes of civic improvement and education, scientific study and conservation, as well as the display of superiority and economic independence.

The first efforts to found a botanic garden in Washington were unsuccessful, and it was not until 1816 that the Columbian Institute for the Promotion of Arts and Sciences was established with the express goal of founding a national botanic garden and a museum of art and natural history. This, the first learned society in the capital, was the direct ancestor of the National Institute and the Smithsonian Institution. The leading figures in intellectual, artistic, and scientific circles in the country—including many medical practitioners, and public figures such the mayors of Washington, members of Congress, and

14. *National Magazine* 1 (October 1801–January 1802): 38.
15. Quoted in Rathburn, "The Columbian Institute," 39.

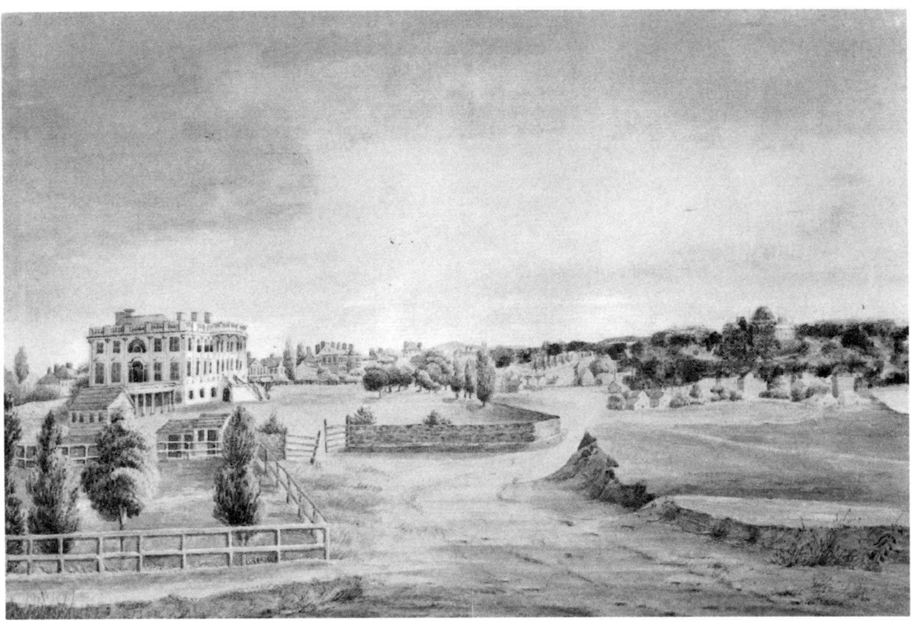

Figure 14. Anthony St. John Baker, *The White House* (1826), watercolor. Huntington Library.

the former presidents—made up a large portion of the membership. Important architects among the members were William Thornton, Charles Bulfinch, and George Hadfield (all designers of the Capitol); James Hoban (architect of the White House); and Robert Mills (architect of the Treasury and Washington Monument). Although there is little concrete evidence of what they did in the institute, it seems likely that John Quincy Adams, a founding member, and Benjamin Henry Latrobe, the first secretary, played key roles in developing the first national botanic gardens. Both were politically powerful figures with scientific acumen, aesthetic sophistication, and commitment to the development of the capital city.

Adams played a critical part in the formation of botanic gardens in this period. Not the first gardener to inhabit the White House, Adams was surely the most ambitious. As president (1825–29) he encouraged the planting of gardens for ornamentation and experimentation and for the preservation of natural resources. He extended the flower beds at the White House to ten thousand square feet and planted literally hundreds of trees and shrubs with his own hands. The grounds were leveled and graded and in time "began to assume a park-like appearance." In a watercolor of the White House by Anthony St. John Baker (1826), part of Adams's arboretum is visible (fig. 14). It contained

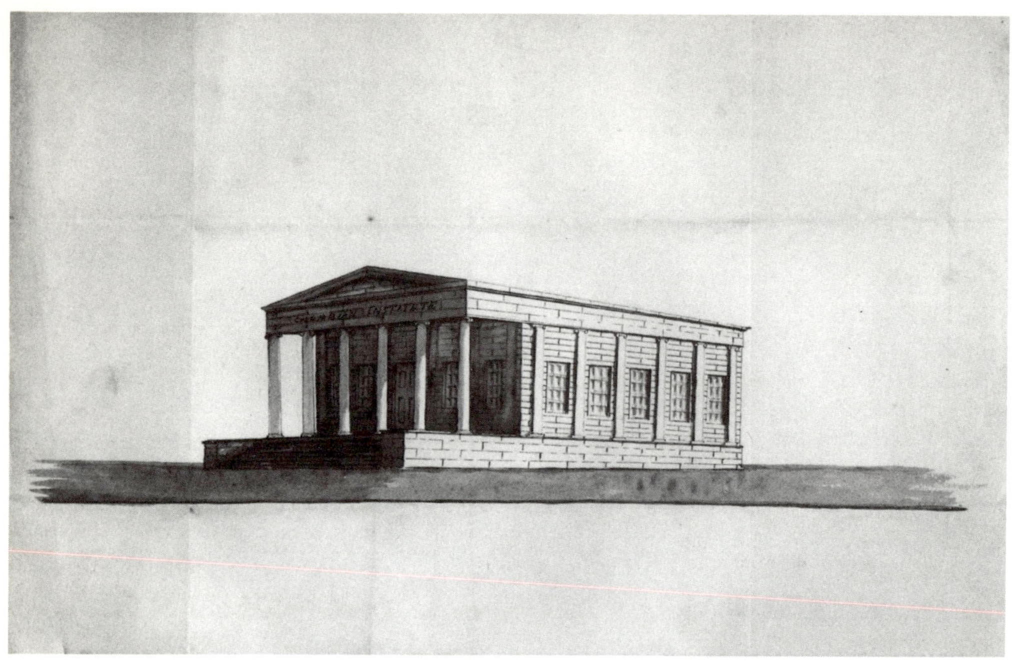

Figure 15. H. Meigs, *Edifice for Columbia Institute* (7 June 1820). Smithsonian Institution Archives, Record Unit 7051, Columbian Institute Records, 1816–1841, and Related Papers, 1791–1800.

hundreds of trees, forming what he claimed was the foundation of an oak and walnut forest that "a century hence may be my successor." An ardent conservationist, Adams sought to preserve the country's hardwood forests for shipbuilding, to identify and save endangered trees, and to introduce foreign species that might be naturalized.[16] Adams not only developed the White House grounds but also paid for the collection and planting of native forest trees growing in the vicinity of the capital for the garden of the Columbian Institute.

In 1818, the Institute appealed to Congress for the appropriation of grounds in order to establish a botanic garden and museum. Five acres were designated at the extreme east end of the Mall between the Capitol and the canal at the foot of Capitol Hill and leased to the Columbian Institute. This was the first lease of public grounds in the still new capital. As members of Congress were all expected to send back native plants from each of their states and districts, the location at the foot of the Capitol provided a symbolic gathering from all parts of the United States (figs. 15, 16).

16. Jack Shepherd, "Seeds of the Presidency: The Capital Schemes of John Quincy Adams," *Horticulture* 30 (1983): 38–47.

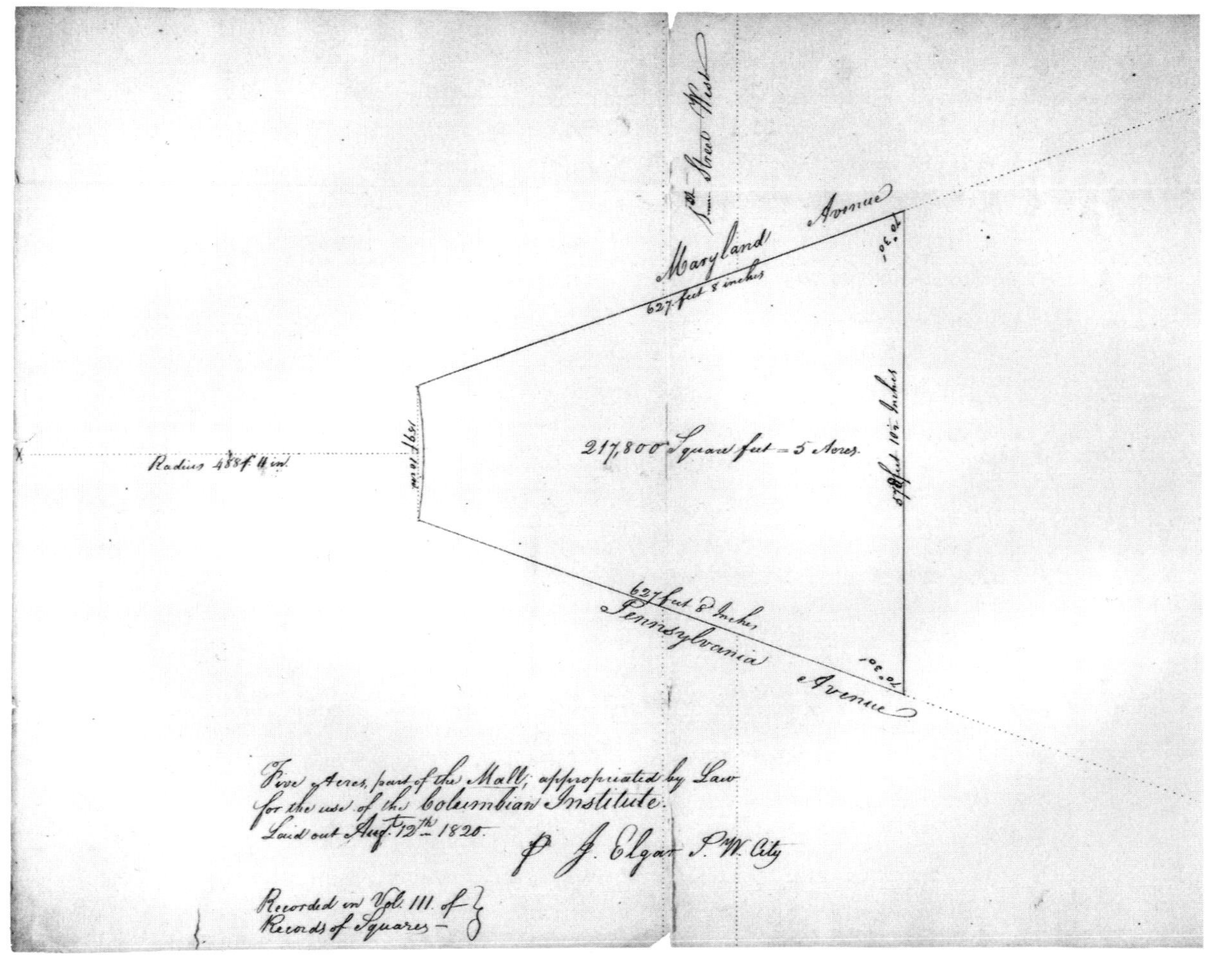

Figure 16. Plan of the Columbian Institute's plot for a botanical garden on the Mall (1820). Smithsonian Institution Archives, Record Unit 7051, Columbian Institute Records, 1816–1841, and Related Papers, 1791–1800.

Improvement of the garden continued, and by 1830 over a thousand shrubs and trees contributed from both domestic and foreign sources had been planted in the garden and reportedly were thriving. The institute had by then also implemented its original goal of distributing throughout the nation seeds and plants cultivated at the garden, a practice continued today by the Department of Agriculture.[17] The gardens of the Columbian Institute

17. This was stipulated in the 1826 Memorial presented to the House of Representatives: "2d. this National Botanic Garden may be used to house all kinds of indigenous and exotic trees, shrubs, roots, grasses, &c., to be distributed to every part of the Union"; see 19th Cong., 1st session, doc. 123, bk. 138, p. 8.

marked the first effort to improve the federal property and ornament the city. This can be seen as the beginning of governmental involvement in landscape and garden design, which evolved in the late nineteenth century into the Corps of Engineers and in the twentieth into the National Park Service.

Through the agency of the Columbian Institute, the Mall had begun to take shape as the ceremonial core of the city. The ambition to create there a utopian "Paradise of America" had not been realized, but an initiative serving educational, scientific, artistic, and civic goals had begun its work. Key support of a financial and political kind would finally come from the federally funded Great Exploring Expedition (1838–42), led by Lt. Charles Wilkes.[18] When the scientists began to send back great quantities of rare and valuable live plant material from all over the world, the necessity of housing it forced the revitalization of the Columbian Institute's gardens, under the aegis of the Joint Committee of Congress on the Library. The gardens then came directly under the guardianship of the federal government. They were to remain on their original site for almost a hundred years.

During the 1840s, the botanic gardens were an important party in the controversy over the Smithson bequest. In 1836, James Smithson, a British mineralogist, left a half million dollars to the United States to be used "to found at Washington, under the name of the Smithsonian Institution, an establishment for the increase and diffusion of knowledge among men."[19] The money remained unused for several years because Congress could not decide what purpose to put it to. Competition for the Smithson bequest among various intellectual, scientific, and cultural groups was keen. Contending proposals included a national observatory, a grand and noble library, and a national university.

In a paper written in 1841, the botanist and physician William Darlington proposed botanic gardens together with a national museum for the Smithson bequest. He urged that it was the duty of national gardens to "procure from every region of the globe perfect specimens of every production of nature." The garden was to represent not only all of nature but also all of knowledge; it was the biblical Tree of Knowledge, gathering up all the "branches" taught in the nation's colleges. The patriotic theme receives a different turn here: the botanic garden was originally supposed to surpass its European counterparts; now it was thought necessary to imitate and equal them. Why, he asked, "when

18. The Great United States Exploring Expedition, often referred to as the Wilkes Expedition, traveled to the polar regions, the South Pacific, the coast of Oregon, Washington, and British Columbia, mapping and collecting scientific specimens of everything encountered. The yield was far greater than anticipated. See William Stanton, *The Great United States Exploring Expedition* (Berkeley, Calif., 1975).

19. G. Brown Goode, "The Genesis of the National Museum," in *Report of the United States National Museum* (Washington, D.C., 1892), 281.

nearly every crowned head in the civilized world had taken care to found such noble institutions as botanic gardens, should not the classic pillars of our Republican fabric be wreathed with the chaplets of Science and festooned with the garlands of taste?"

> And while the Frenchman justly glories in the Jardin des Plantes—while the Briton boasts with reason, of the royal Garden at Kew; and even the russian, in his frozen clime is warned in admiration of the Imperial Conservatory of the Czars—let American freemen, in their turn be enabled to point with patriotic pride, to a National institution of no less beauty and value, at the Metropolis of their own favored land. While at colleges they teach the various branches of knowledge, here at the common center of the Republic, we should have the entire Tree, in perennial verdure, accessible to all who might desire to participate in its pleasures and benefits.[20]

In the same year, Joel Poinsett, secretary of war, ordered that the collections made by the Wilkes Expedition, which were being sent all around the country, be returned to Washington. He convened a group of intellectuals to establish a "cabinet of natural history," a museum and repository for these natural history curiosities. This group created the National Institution for the Promotion of Science with the hope of receiving the Smithson bequest and of achieving the goals that the Columbian Institute had originally envisioned.[21] Robert Mills, federal architect, was asked to make the plans.

Poinsett's concerns were in fact much broader than the protection of the Wilkes material. Unlike earlier proponents of botanic gardens, he was aware of the threat to native flora that resulted from national industrial development. The role of the gardens would be

> to collect documents and facts illustrative of the early history of our country, specimens of its geology and of its mineral and vegetable productions, and if not to preserve the animals and plants themselves, which are passing away before the progress of settlement and cultivation, at least to perpetuate their forms and the memory of their existence.[22]

20. William Darlington, *A Plea for a National Museum and Botanic Garden to be founded at The Smithsonian Institution at the City of Washington* (Westchester, Pa., 1841), 10–12.

21. Goode, "Genesis of the National Museum," 288.

22. Ibid., 290.

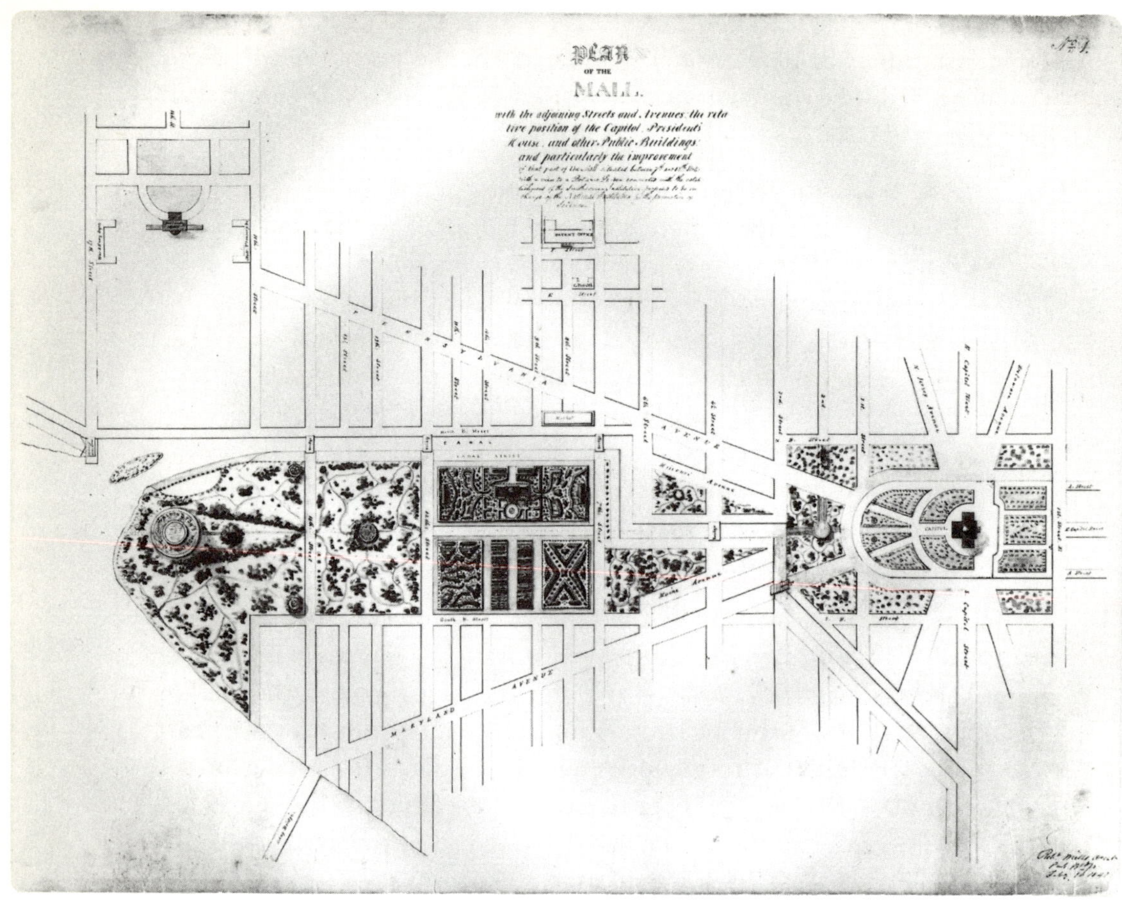

Figure 17. Robert Mills, Plan of the Mall (1841). Civil Works File, Cons. 90.1, Office of the Chief of Engineers; Record Group 77, National Archives, Washington, D.C.

By the early 1840s the Mall was sorely in need of improvement if it was to set a standard for other American cities, as was appropriate to the seat of the federal government. It was generally realized that Washington, despite its location along a primary route to the west, could not compete economically with New York, Boston, or Philadelphia. But it could take the lead in art and science, and in symbolic stature. This sentiment, along with the momentum generated by the Smithson bequest, moved Congress finally attend to the National Mall.

Mills proposed a picturesque assembly of different gardens extending from his Washington Monument to the Capitol. The most complicated part of the plan was for the botanic gardens surrounding the museum building (figs. 17, 18, 19). But the gardens would be picturesque only in aggregate;

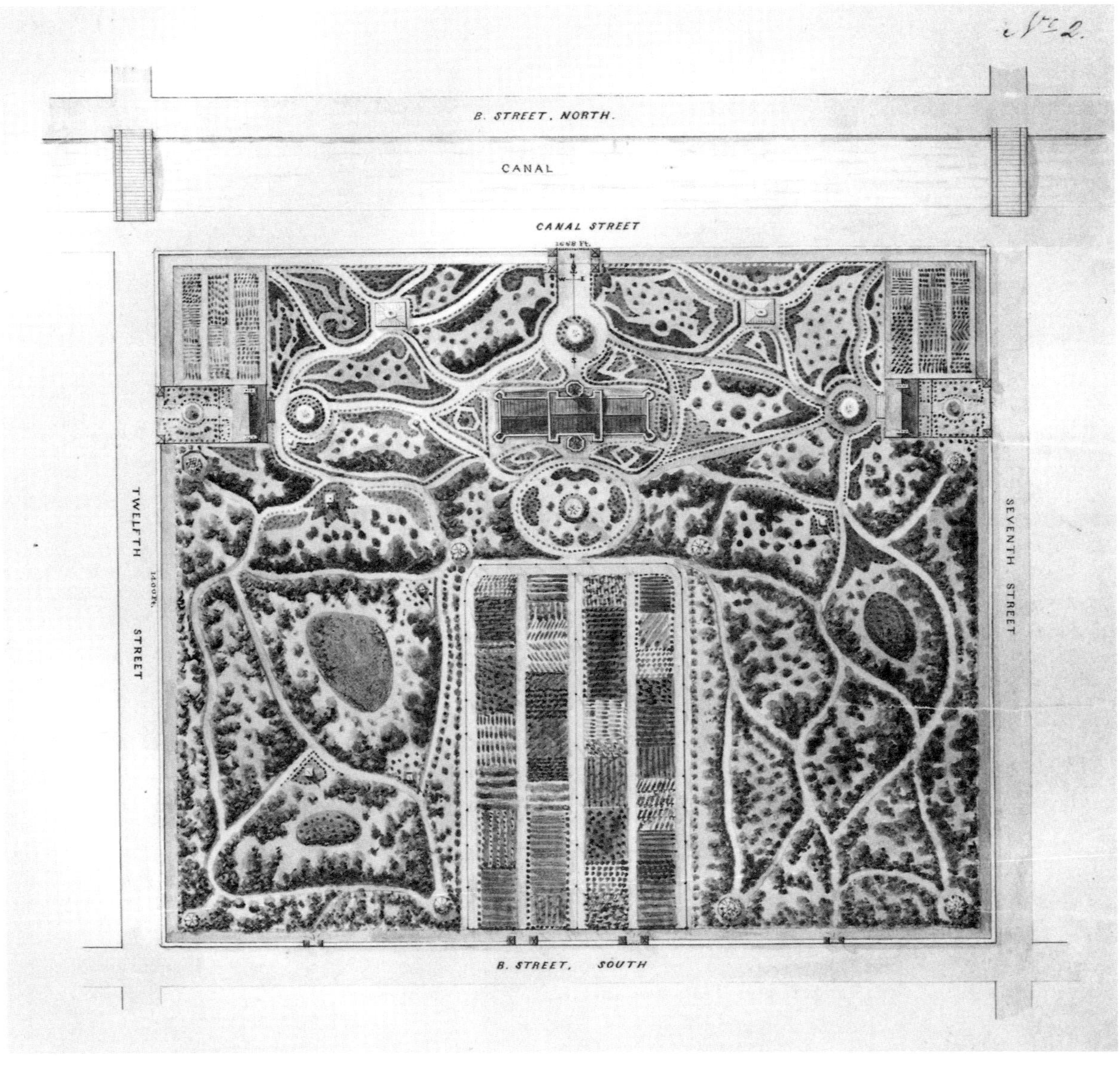

Figure 18. Robert Mills, Alternative plan for grounds for the Smithsonian Institution (1841). Civil Works File, Cons. 1.3, Office of the Chief of Engineers; Record Group 77, National Archives, Washington, D.C.

Mills seems to have been strongly influenced by a new and more overtly horticultural design called the gardenesque, characterized by highly controlled planting and quite distinct from the earlier picturesque style. This approach reflected the new emphasis on the botanic and practical aspects of gardening. It had become more dominant since the late eighteenth century, and it

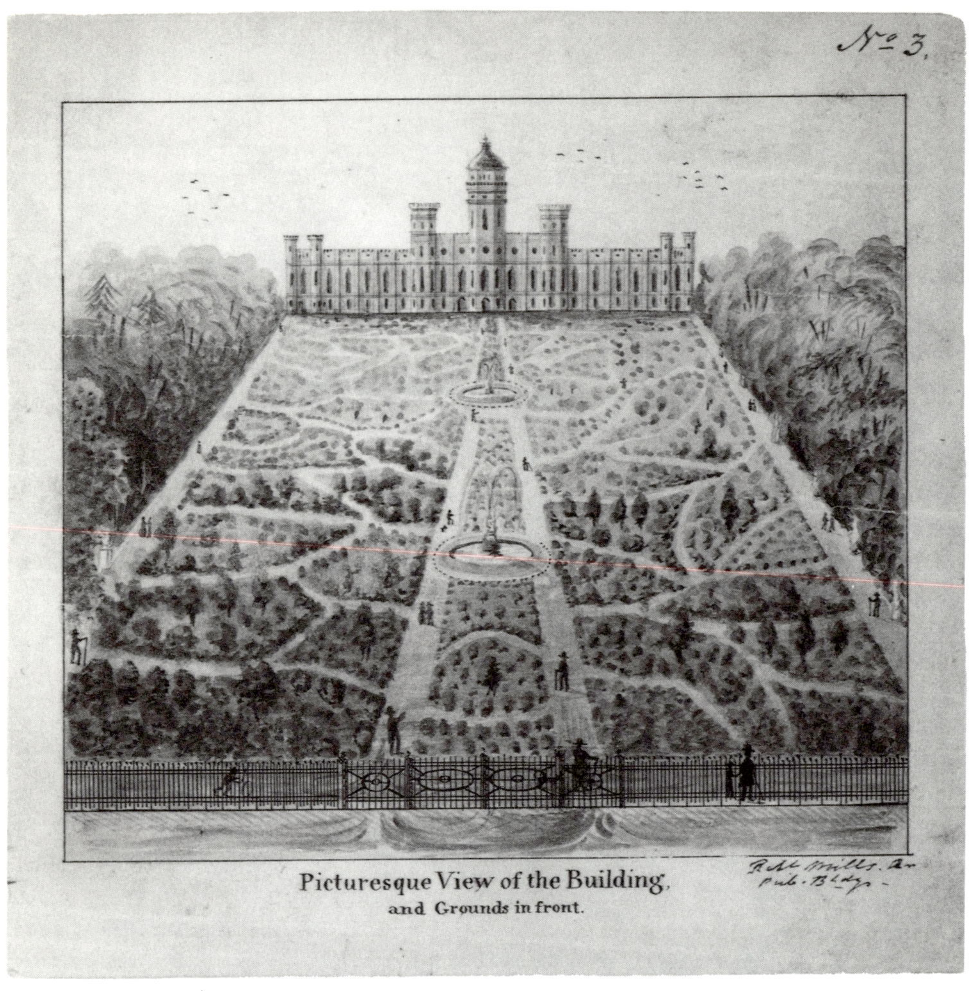

Figure 19. Robert Mills, Picturesque view of the Smithsonian Institution building and grounds in front (1841). Cons. 1.2, Record Group 77, National Archives, Washington, D.C.

was formalized by John Claudius Loudon, who designed the Birmingham botanic garden in 1831 and the Derby arboretum in 1839.[23] Loudon recommended using the "natural system" of Jussieu, which had by the 1830s replaced the Linnaean sexual system of classification. In the gardenesque mode, plants were separated according to the natural system—that is,

23. A copy of Loudon's *Encyclopedia* (1824) that Mills may have seen is in the holdings of the Library of Congress; an inscription on the inside front cover reads "Department of the Treasury," the office in which Mills worked as architect of public buildings.

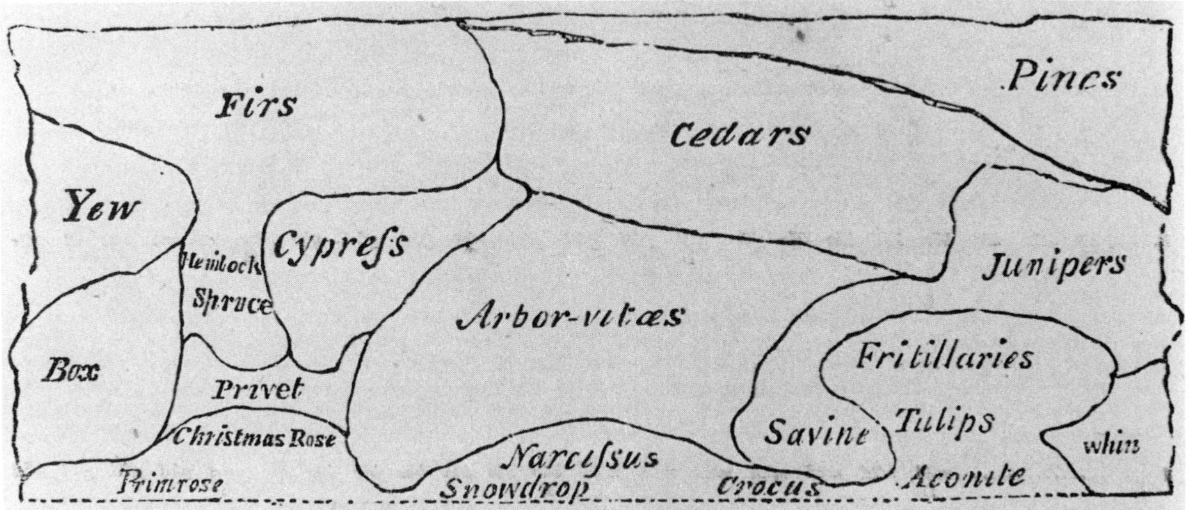

Figure 20. J. C. Loudon, The select or grouped manner of planting shrubbery, from *The Encyclopedia of Gardening* (London, 1824). Library of Congress.

according to the greatest number of similarities in form (fig. 20).[24] Mills's drawings show an articulation of individual beds, suggesting his use of the gardenesque style and the natural system.

Mills created an architectural and landscape setting in a style that had not yet been used for public projects in Washington and one that would set the museum compound apart from the rest of the city. He drew on European models in his creation of a campus of academic and residential buildings linked by botanic gardens and enclosed by a wall—not unlike a medieval monastic community. The proposed scientific gardens of the National Institution, furthermore, alluded to such botanic or physic gardens as those at Oxford and Chelsea. The English medieval styles were associated with institutions of learning and, of course, with Smithson's nationality.[25]

24. In the *Suburban Gardener and Villa Companion* (London, 1838), Loudon defined the gardenesque style (pp. 164–66) as "the production of that kind of scenery which is best calculated to display the individual beauty of trees, shrubs and plants in a state of nature; the smoothness and greeness of lawns; and the smooth surfaces, curved directions, a dryness and firmness of gravel walks; in short, it is calculated for displaying the art of the gardener."

25. Mills, in a letter to Robert Dale Owen entitled "A Report on the Proposed Smithsonian Institute," stated that "Our Associations with great literary institutions are assimilated with the Saxon style of architecture in those buildings, and we may presume, that were the honored Mr. Smithson living at this time, and were funding such an institution, he would give it preference"; see H. M. Pierce Gallagher, *Robert Mills: Architect of the Washington Monument, 1781–1858* (New York, 1935), 193.

The museum and gardens in Mills's design for the Mall should be understood as part of a single composition that also included a medieval museum, the Greco-Egyptian Monument, and various bridges, temples, and fountains that functioned as garden structures within a varied landscape. Thus his adherence to the gardenesque was situated in the older tradition of the picturesque. Such a melange of architectural and garden types also had older European roots, in picturesque landscape parks and in the royal botanic gardens at Kew, designed by William Chambers.

Mills's design can also be understood as part of an ambition to create a city-wide "museum" of architecture and taste in Washington. It has been suggested that Mills had an iconographic program for the whole city, which he intended to transform into a museum of references to world architecture, linked by gardens of reference and association.[26] This would have been consistent with the general movement during this period to make Washington a "City of Reference."[27] It was also reminiscent of Thomas Jefferson's didactic intention as reflected in the architecture of the University of Virginia, which comprised a series of pavilions displaying classical orders of knowledge for the instruction of students.

Continued efforts to improve the Mall were surrounded by political controversy. The use of the Mall for agricultural and botanic gardens under the control of the Smithsonian was opposed chiefly in the deep South, where it was felt that the multifaceted institution was an inappropriate expenditure of federal funds, an extension of federal powers into areas better left to each state. Despite the Mall's unimproved condition at this time, the controversy made clear its potential power as a symbol of the federal government. After a series of stalls and compromises, Mills's plan for the Mall was finally set aside in favor of one by James Renwick. The Mall was not sited in a botanic garden but in an arboretum of indigenous trees.[28]

❧ ❧

Although Mills's work had opened discussion on the role of the Mall as a showcase of American scientific and artistic achievement, in the mid-nineteenth century it was still mostly unimproved. A visitor wrote a poignant description at

26. Pamela Scott, "'This Vast Empire': The Iconography of the Public Space of the Mall, 1791 to 1848," in Richard Longstreth, *The Mall in Washington* (Washington, D.C., 1991), 37–58.
27. Rathburn, "The Columbian Institution," 54
28. Smithsonian Institution, "Annual Report of the Board of Regents of the Smithsonian Institution, 1846," in *Senate Executive Documents,* 29th Cong., 2d sess., Doc. 211, esp. p. 29; and Smithsonian "Annual Report . . . 1847," 83.

the time of the half-built monuments and jumbled treasures of the various natural history and art collections locked in temporary quarters. Here he found a metaphor for the undeveloped potential of the still young nation, which had awakened in him "that painful sense of the incomplete or that almost perplexing consciousness of the new, the progressive, and the unattained which is peculiar to our country."[29] The capital had to fight its reputation not only as unimproved architecturally but also as a center of slavery. In 1850, legislation—the so-called Great Compromise—stipulated that slave-dealing be banned in the Capitol. Slave pens, which had long been the focus of criticism by abolitionists, were finally removed from the Mall.

The antebellum period thus saw an increased effort to construct gardens in the capital that could be considered "democratic" as well as ornamental. Launched despite the fact that the nation was on the threshold of a civil war, or perhaps because of it, this effort was intended to convey the image of a unified republican government committed to popular education and a democratic society. Congress, with unprecedented federal spending, once again attempted to develop the Mall as the ceremonial core of the city.[30]

It was the pedagogical function that had always been linked to gardens that was especially attended to in the Mall's development as a national park. This was part of a nationwide proliferation of botanic gardens, museums, public libraries, and free schools that resulted from the movement for popular education that swept the country during the 1840s and 1850s. A future program for the Mall was depicted allegorically in Bruff's drawing of 1845, *Elements of National Thrift and Empire* (fig. 21). Within the next five years, the Washington Monument, the Smithsonian, the city canal, and a landscape park were all incorporated into a single plan for the national capital, a single powerful statement of a political and educational vision.

In 1850 President Millard Fillmore asked Andrew Jackson Downing of New York, the country's preeminent landscape gardener and an advocate of agrarian virtue, to develop a plan for the Mall to include the new Smithsonian Institution grounds. Downing, the "rural architect," was the obvious choice for this commission. He had been championing public parks for American cities for many years. Through his magazine *The Horticulturist* and his enormously popular books on landscape and architectural theory, Downing had almost single-handedly stimulated an unprecedented interest in and knowledge of the art of garden design among Americans. He was furthermore

29. H. T. Tuckerman, "The Inauguration," *The Literary Messenger,* no. 15, pp. 236–40, quoted in Goode, 329.
30. Constance McLaughlin Green, *A History of the Capital, 1800–1950* (Princeton, N.J., 1962), 180, 199.

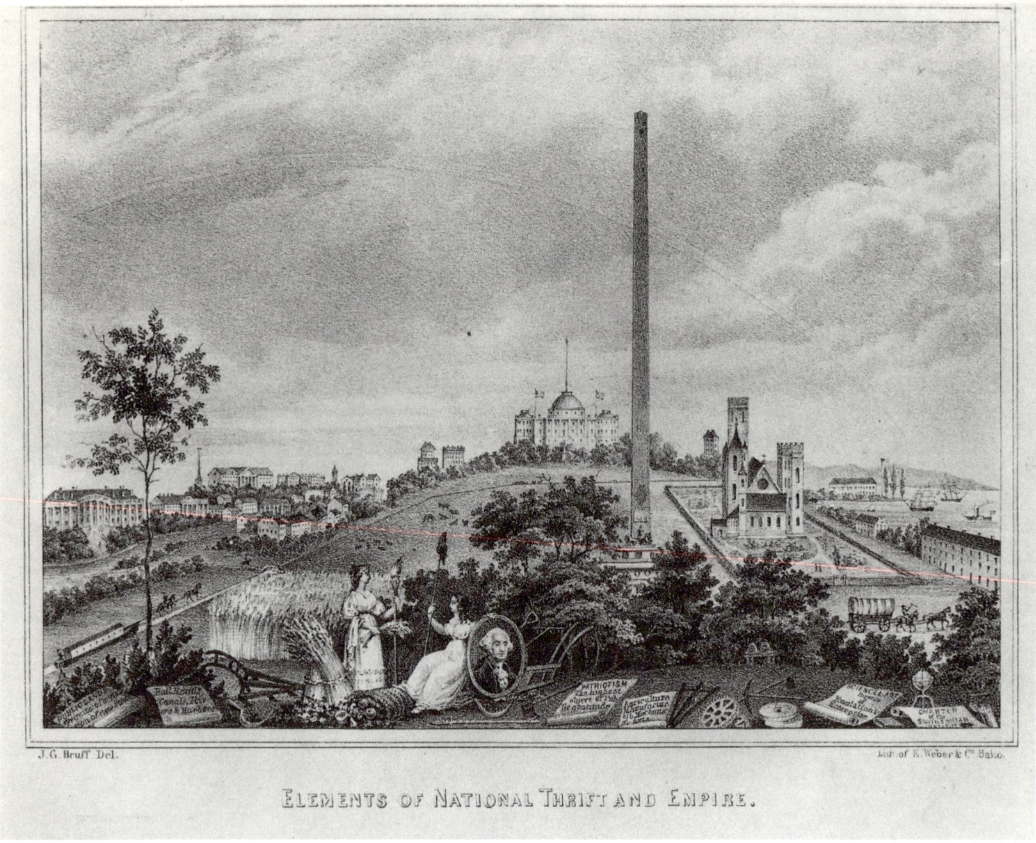

Figure 21. E. Weber, after drawing by J. G. Bruff, *Elements of National Thrift and Empire* (1847). Library of Congress.

recognized internationally as the arbiter of American taste.[31] In terms of landscape gardening, his plan advocated a "natural style." The eyes of the nation, in this plan, were turned less on the Old World and more toward the new American states and territories to the west.

Downing's plan for laying out the public grounds connected the Capitol, Smithsonian, and White House grounds by means of a series of six different but compatible gardens (fig. 22). He described his intentions in a letter to President Fillmore:

31. The high esteem in which Downing was held was expressed in Congress by Senator Bell just after Downing's death: "I understand he had no equal in this country in that line. I do not pretend to be any judge of such subjects myself; but I regard his loss as a public calamity, as has been said by honorable Senators. Such things as rural architecture: the improvement of ground around your cottages, if you please—because the laying off of the small grounds around the cottage is a part of the wealth even of the poor cottage of the country—are important" (*The Congressional Globe* [26 August 1852], 2375).

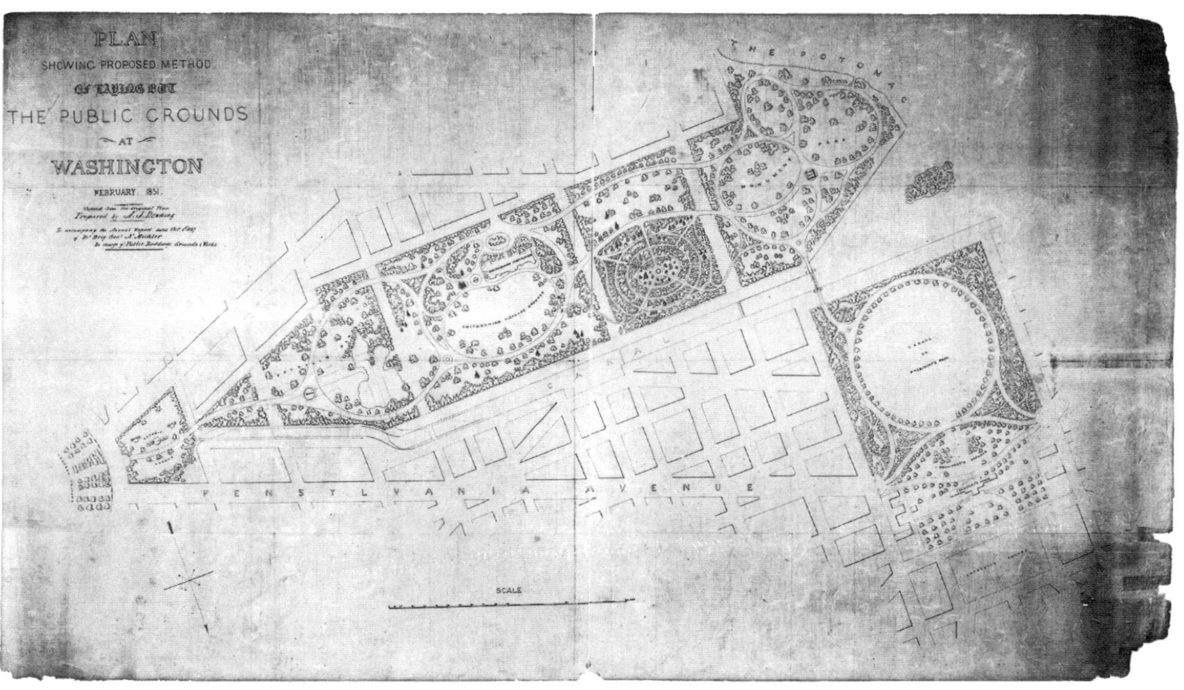

Figure 22. J. Downing, Plan showing proposed method of laying out the public grounds at Washington (1851), copy by N. Michler (1867). Record Group 77, National Archives, Washington, D.C.

My object is threefold. First, to form a national park which would be an ornament to the Capital of the United States; 2. To give an example of the natural style of landscape gardening which may have an influence on the general taste of the Country; 3rd: To form a collection of all the trees that will grow in the climate of Washington, and, by having these trees plainly labelled with their popular and scientific names, to form a public museum of living trees and shrubs where every person visiting Washington would become familiar with the habits and growth of all the hardy trees.[32]

In Downing's plan, the old botanic garden and greenhouses were expanded to accommodate the collections brought back from the Wilkes Exploring Expedition. Thus these grounds at the foot of Capitol Hill initiated what could

32. Andrew Jackson Downing, "Explanatory Notes to Accompany the Plan for Improving the Public Grounds at Washington"; National Archives, Record Group 42, LR, vol. 32, no. 1358 1/2, 3 March 1851. For a discussion of Downing's plan see Therese O'Malley, "'A Public Museum of Trees': Mid-Nineteenth-Century Plans for the Mall," in Longstreth, *The Mall in Washington*, 61–76.

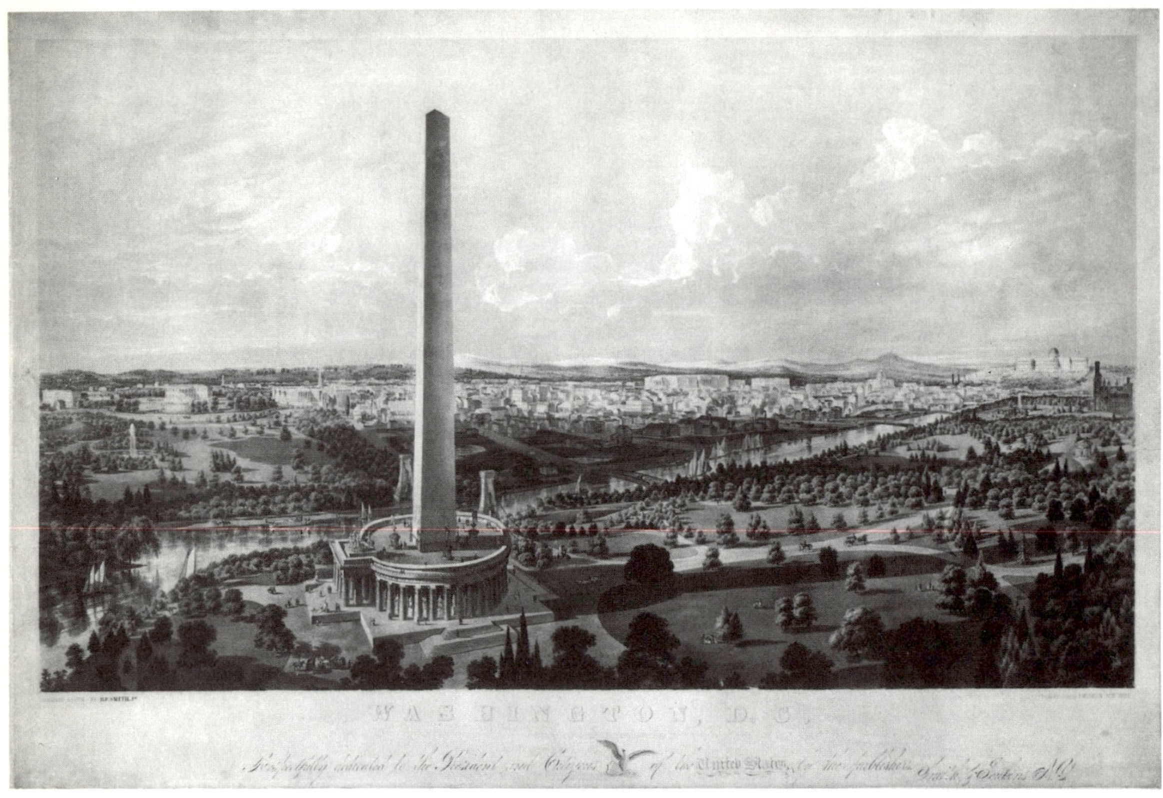

Figure 23. B. S. Smith Jr., *Washington, D.C., with projected improvements* (1842), lithograph. Library of Congress.

be conceived as a physical representation of the Union: the federal government, represented in terms of its native trees, connected to the recently annexed transcontinental states and territories, represented in terms of specimens gathered from all of them (fig. 23).[33]

Downing was heir to one hundred years of botanic gardening in America. It was a century, in the nation's capital in particular, that experienced both a growing interest in the American natural environment and a shift in sensibilities about man's relation to it. What took place in Washington is a significant index of garden history in this country because the Mall, a self-conscious American icon, was the focus of every phase of that history. The Mall received the attention not only of scientists and architects but also of the major thinkers and

33. In 1848, the Treaty of Guadalupe-Hidalgo was signed, adding a large piece of the present-day west and southwest, making the United States transcontinental; see Green, *History of the Capital,* 177.

political leaders during a period in which the nation was trying to establish itself in the eyes of the world. To see the persistent effort to develop botanic gardens on the Mall in Washington is to become aware of the American obsession with the idea of gardens, and with their capacity to express political and aesthetic ideas as well as attitudes toward the past and hopes for the future. The early history of the Mall is a story of driving aspirations and repeated attempts to create a physical environment that celebrated the natural gifts, cultural acumen, and political vitality of the nation. The early American botanic gardens in general, and those on the Mall specifically, embodied the fundamental belief in the perfectibility of man and the optimism of the founders in the future they were creating for the new republic.

National Gallery of Art

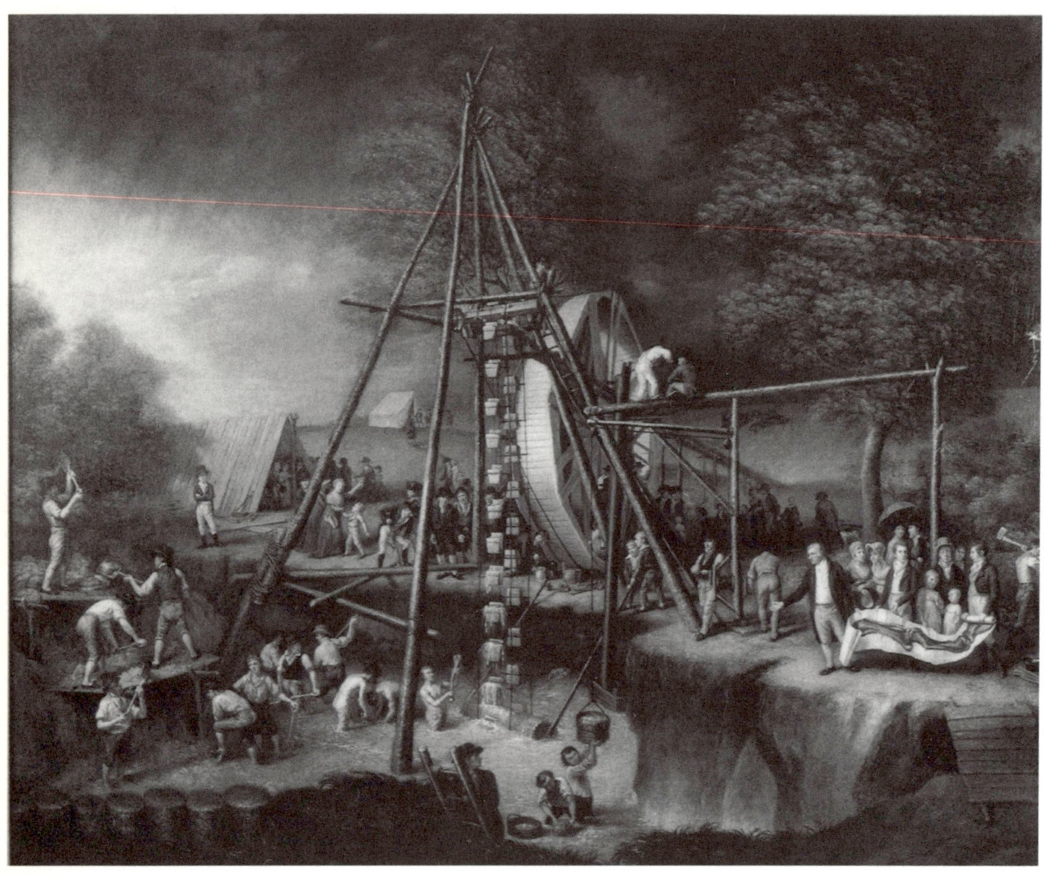

Figure 24. Charles Willson Peale, *Exhumation of the Mastodon* (1807); oil on canvas, 127 x 158.8 cm. Maryland Historical Society.

Empire of Birds:
Alexander Wilson's *American Ornithology*

Laura Rigal

> I spent nearly the whole of Saturday in Newark, where my
> book attracted as many starers as a bear or a mammoth would
> have done.
>
> —Alexander Wilson[1]

In 1807, Charles Willson Peale included a portrait of Alexander Wilson in his painting *Exhumation of the Mastodon* (fig. 24). In the painting Wilson stands, arms crossed, above a line of laborers who are digging for bones below. In a painting full of pyramidal structures—from the tent in the background, to the shack in the middle-ground, to the machine looming in the foreground—Wilson is positioned at the peak, or pinnacle, of the laborers' progressive work. Gazing down upon them, he figures in his contemplative stance the end and product of their labor. Among the artifacts of the mammoth excavation, it seems, are not only the restored bones of the mastodon but also this solitary watcher. The summa of productive labor, or so Peale's painting reads, is this (representation of a) representative man—Alexander Wilson, the author of the first American ornithology, in whose reflective pose work is joined with leisure, labor with spectatorship. This essay considers the great labor of literary production that Peale so admired: Wilson's illustrated *American Ornithology*, published in nine volumes by Bradford and Inskeep of Philadelphia between 1807 and 1814.

Like Peale's museum and its mammoth exhibit of 1801–10, Wilson's collection of some five hundred bird species was an artifact of productive labor

This essay owes much to the help, encouragement, and advice of Lauren Berlant, Zofia Burr, James Chandler, Richard Horwitz, Janice Knight, Christopher Looby, Trish Loughran, Jonathan Sachs, Katie Trumpener, and William Veeder; and to the editorial skills of Trish Loughran. It was also improved by comments offered by David Brigham, Amy Meyers, Robert C. Ritchie, and Alan Taylor at the Huntington conference where it was first delivered in 1994.

1. Letter to a friend, 10 October 1808, *The Life and Letters of Alexander Wilson*, ed. Clark Hunter (Philadelphia, 1983), 276.

earnestly and eagerly performed. It was also, however, an artifact of western expansion and exploration that displayed new western species from the continental interior (Tennessee, Mississippi, and the Louisiana territory) together with species long familiar on the east coast (blue jay, robin, oriole, bluebird, and so on). Like Lewis and Clark, Wilson took the exhibitionary practices of Jeffersonian science on the road, bringing the representational devices of the Peale museum to the fields and streams of the western interior. At the same time, Wilson relied heavily on Peale's Philadelphia collection of more than seven hundred bird species, mounted and exhibited in the upper floor of Independence Hall (then the Pennsylvania State House). A quintessentially federalist work, Wilson's *Ornithology* unites myriad local specimens within an explicitly national collection. But, even more centrally, as a mammoth performance of production, *American Ornithology* joins labor with looking, visual with verbal genres, framing its readers and viewers as producers and spectators—by exhorting them to learn to *see* federalist-style, or from the elevated (and diffused) point of view that would constitute them as independent, self-regulating subjects and objects of the state.

In his essay "On Manufactures" in *Notes on the State of Virginia,* Thomas Jefferson celebrated the American husbandman, or agricultural producer, as the source of collective independence and virtue. It is the husbandman, Jefferson writes, who "keeps alive th[e] sacred fire" of virtue "which otherwise might escape from the face of the Earth." Relying upon his own labor rather than upon "the caprice of customers," the farmer is the privileged working man of Jefferson's landscape: "Corruption of morals in the mass of cultivators is a phaenomenon of which no age nor nation has furnished an example."[2] In this classic statement, Jefferson grounded independent nationhood on the virtuous labor of the husbandman. But, in fact, Jefferson's agrarian republic was never merely the product of agricultural labor; it also depended crucially upon graphic and visual art to frame the landscape of solitary producers that, he imagined, would preserve national independence. And this is what the illustrated text of Wilson's *Ornithology* brings home: the visual as well as verbal structures of representation (the techniques of collection classification, publication, illustration, and museum display)—by which Jeffersonian federalism construed its subjects not only as productive workers, or assembly-men, but as viewers and spectators as well. It is here, then, where art and science converge in the framing of human subjects, furthermore, that we must consider the arts and sciences not only as modes of representation, but as technologies, or means

---

2. Jefferson, "On Manufactures," *Notes on the State of Virginia,* ed. William Peden (New York, 1954), 165.

of production. It is thus that we find in the visual and linguistic forms of Jeffersonian science that we find the deep structure of American federalism as a species of empire.

As early as 1786, Jefferson himself had described the thirteen former colonies as a "nest" from which the entire continent would be peopled. By the mid-1780s, emigration west (to Kentucky and Ohio in particular) had raised fears that the confederation was growing too fast and would simply fragment as its political authority extended over too great an area. In the following passage, Jefferson tries to counter these fears by figuring the Anglo-American emigration to Kentucky as natural extensions of a national "nest" through a peaceful diffusion of settlement comparable to a migration of birds:

> Our present federal limits are not too large for good govern-
> ment, nor will the increase of votes in Congress produce any ill
> effect. . . . Our confederacy must be viewed as the nest from
> which all America, North and South is to be peopled. We
> should take care not to think it for the interest of that great
> continent to press too soon on the Spaniards. Those countries
> cannot be in better hands. My fear is that they are too feeble to
> hold them till our population can be sufficiently advanced to
> gain it from them piece by piece.[3]

Here, bird migration is a metaphor for the peaceful and gradual replacement of a "feeble" Spanish empire by settlers moving west in the form, presumably, of Anglo-American freeholders or husbandmen. Together with his idealized yeoman farmer, Jefferson's metaphor of the former colonies as a nest is his most telling figure for territorial expansion. The nest not only harmonizes a familial subject with agricultural production; it also naturalizes the grid of property formation. Jefferson's husbandman was always an agent of expansion; as Jefferson himself put it, "we have an immensity of land courting the industry of the husbandman."[4] But, as the metaphor of the nest implies, the establishment of (this state of) cultivation depended not merely upon the mobility of Anglo-American farmers but also upon the containment and rationalization of their labor by art and science. Here, where agrarian political economy is figured by the nest, cultivation becomes acculturation, and the family circle of Jeffersonian fantasy is revealed as a device for the containment as well as the diffusion of productive labor.

---

3.   Jefferson to Archibald Stuart, 25 January 1786, *The Papers of Thomas Jefferson*, vol. 9, ed. Julian P. Boyd (Princeton, N.J., 1954), 217–18.
4.   Jefferson, "On Manufactures," 164.

It is this metaphor of the nest—of union as a phenomenon of production and its migratory reproduction—that Alexander Wilson's bird books carry to its logical republican extension. Hauling his sample volume of *American Ornithology* from city to city, from settlement to settlement, in search of new subscribers and new bird specimens, Wilson traveled through virtually every city "from the shores of St. Laurence, to the mouths of the Mississippi," from the Atlantic ocean to the interior of Louisiana. He described his journeys through the United States and the territories of the interior as a "zigzag" from "one country to another."[5] As a result, however, his nine-volume collection of the "manners and migrations" of American birds is something more than an oversized field guide to the birds. It is, instead, a fragmentary kind of panorama, unrolling the expansionist inner state of production's reproduction during the years of Jefferson's and Madison's presidencies (1801–16).

Wilson's *American Ornithology* was, above all, a collective commodity—a self-consciously all-American assembly and display of United States materials and productive power. As its author proclaims in the preface to his fifth volume, its "engravings are a monument to the merits of Messr's Lawson, Murray and Warnicke, the elegance of the letter press a high honor to the taste of the founders Binney and Ronaldson, . . . while the paper, from the manufactory of Mr. Ames, proves what American ingenuity is capable of producing when properly encouraged."[6] The colors of the plates were inconsistent with this objective, however. Although all the other "materials and mechanical parts of this publication have been the production of the United States," the author was "principally indebted to Europe" for his colors—except for the "beautiful native ochres" mixed in "the laboratory of Messr's Peale and Son." However, Wilson adds, "the spirit for manufactories, every day rising around us," offers hope that American artists and authors will shortly be rendered "completely independent of all foreign aid" and enabled "to exhibit the native hues of his subjects in colors of our own, equal in brilliancy, durability and effects to any others."[7]

The appropriation by the United States of the continent's midsection is still popularly recounted as a historical narrative, according to which exploration of the western interior and incipient settlement (by scouts, traders, and "white savages") led sequentially to general immigration, cultivation, and domestication. According to this narrative, the interior was first penetrated by

---

5.   Alexander Wilson, *American Ornithology,* 14 vols. (Philadelphia, 1808–14), vol. 1 (1808), introduction, 1–2. All references to Wilson's prefaces are to this first edition of *American Ornithology.* All other references are to Sir William Jardine's three-volume, "unabridged" edition (which omits Wilson's prefaces): Alexander Wilson and Prince Charles-Lucien Bonaparte, *American Ornithology, or The Natural History of the Birds of the United States* (London, 1876).

6.   Preface, *American Ornithology,* vol. 5 (1812), x.

7.   Ibid., vol. 2 (1810), vi.

solitary white men who fought, traded, and often intermarried with Native Americans, and only then were followed by families of homesteaders and, finally, by the institutions and artifacts of national culture: towns, laws, churches, commerce, and consumer goods.[8] Wilson's *Ornithology*, however, makes it clear that, on the contrary, these imaginary processes were not narratively or temporally sequential but simultaneous: territorial expansion was always inseparable from the production, distribution, and marketing of consumer goods—from shoes, clothing, hats, and guns, to books, magazines, and pictures—by an American manufactory that produced not only things but also persons (including the Alexander Wilson of Peale's *Exhumation of the Mastodon*) as representative commodities.

There is no question that Wilson aimed to bring himself forth from obscurity, together with the birds he classified, by virtue of his enormous labor of collection and assemblage. In doing so he embraced, like Peale, a Jeffersonian plan of union through exemplary self-production or yeoman independence within the eye of the state; in the process, he also extended the federal republic with its "elevated" views (that is, disinterested, representative, virtuously synthetic) and its reproductions of production, into the interior. With his mammoth *American Ornithology*, Wilson took the federal roof on the road, as an exemplar of its representational principles and practices. In doing so, however, he also revealed the sharp, violent, and ragged limits of federalism's applicability—the limits, that is, of the productivist principles of American manufacturing as the frame and fabric of a universal, natural history of the world.

Wilson's *Ornithology* was only one of a number of enormously ambitious, expensive, multivolume publishing projects that were undertaken in Philadelphia at about the same time. In his address to the new Pennsylvania Academy of Fine Arts in 1810, Joseph Hopkinson, the son of federalist poet Francis Hopkinson, noted the remarkable growth of Philadelphia publishing since 1776. According to Hopkinson, the number of engravers in the city had increased from three to sixty since the American Revolution, owing to a remarkable boom in book-making.[9] By 1810, the labor of virtually every

8. George Dekker has provided the best account to date of this tradition in *The American Historical Romance* (Cambridge, 1987), 73–98.

9. Joseph Hopkinson, "First Annual Discourse to the Pennsylvania Academy of Fine Arts" (Philadelphia, 1810), 15–16. The Academy of Fine Arts would serve as both a school and a professional organization for American engravers. Most Philadelphia engravers became members of the Pennsylvania Academy, where some, such as the silver engraver/trompe l'oeil painter William Harnett (1848–92), would also find training in other arts. By contrast with the Pennsylvania Academy, the British Royal Society of Sir Joshua Reynolds established itself as an academy of fine arts by virtue of its exclusion of (mechanical) engravers. See Alfred Frankenstein *After the Hunt: William Harnett and Other American Still Life Painters, 1870–1900* (Berkeley and Los Angeles, 1953), 29–33.

engraver active in the United States had been required by one or another of the publishing projects under way in Philadelphia: in 1790 Thomas Dobson began to print the first American edition of the *Encyclopedia Britannica,* which, by 1803, would be completed in twenty-one volumes with six hundred copperplate engravings; between 1795 and 1796, Bioren and Madan of Philadelphia published the first complete American edition of Shakespeare in eight volumes (probably edited by Joseph Hopkinson); and, in 1798, William and Thomas Birch began to engrave their series of twenty-eight *Views of Philadelphia,* published in 1800 for subscribers in Philadelphia, New York, and Baltimore. But even these ambitious projects were dwarfed by the American edition of Abraham Rees's *Cyclopedia, or Universal Dictionary of Arts, Sciences, and Literature,* published between 1810 and 1827 by Samuel Bradford (with Wilson's help as agent) in forty-seven volumes, with over fourteen hundred engravings.[10]

Bradford and his New York partner hired Wilson as book agent and assistant editor to the *Cyclopedia* and paid him nine hundred dollars a year. In so doing they piggybacked the publication of Wilson's *American Ornithology* onto the larger project of the *Cyclopedia,* enabling Wilson to travel across the continent in search of subscribers and information not only for the *Cyclopedia* but also for his own projected *Ornithology.* Travel was critical to the completion of both projects, not only for the purposes of collecting specimens and data but also in order to contact local bookdealers, and to scour the continent for subscribers wealthy enough, and willing, to capitalize such expensive commodities.

Illustrated with almost one hundred plates (drawn by Wilson, engraved by Alexander Lawson, and hand colored by men and women hired by Wilson), Bradford and Inskeep's edition of *American Ornithology* was in every sense a luxury commodity; its price of $120.00 (or $12.00 per volume) was prohibitive for all but the wealthiest individuals, and institutions such as library companies, learned societies, and colleges. Wilson's subscription list, published in the ninth volume of the *Ornithology,* crosses partisan lines, replicating, as a list of names, the portraits of republican worthies that lined the wall of the Long Room in Peale's Museum (see figure 50 in Linda Partridge's essay in this volume). The list exhibits the names of New England federalists Jedediah Morse and Josiah Quincy, Virginians Thomas Jefferson and James Madison, as well as Thomas Pinckney, Nicholas Biddle, Benjamin Smith Barton, Rufus King,

---

10. See David McNeely Stauffer, *American Engravers upon Copper and Steel* (New York, 1907), xxvii. See also Mantle Fielding's *Supplement to Stauffer's American Engravers* (Philadelphia, 1917); Ellis Paxson Oberholtzer, *The Literary History of Philadelphia* (Philadelphia, 1906), 149; and John Tebbel, *A History of Publishing in the United States: The Creation of an Industry,* vol. 1, 1630–1865 (New York, 1972), 174.

Robert Livingston, and Gouverneur Morris.[11] Such elite individuals chiefly comprised the subscribership of 448, broadened only slightly by institutions: eleven city libraries, five learned societies, six universities, an agricultural society, a state legislature, a hospital, and a Scottish Museum.

Sharply restricted in terms of class and profession, Wilson's subscribers constitute a group that was, nevertheless, geographically extensive, assembled by a Philadelphia publishing firm with expanding, or federal, marketing range. Prior to the canals and turnpikes of the 1820s, publishing was, like so many American enterprises, a profoundly local affair. Only Philadelphia and New York could claim a growing book trade with the southern and western interior.[12] The *Ornithology's* subscribers represent Pennsylvania (with 114 names), New York (60), Connecticut (3), Massachusetts (14), New Hampshire (8), Maryland (16), District of Columbia (20), Virginia (34), North Carolina (11), South Carolina (36), Georgia (21), Kentucky (15), Mississippi Territory (23), Louisiana (60), and finally "Europe," with 15 subscribers (from England, including London and Liverpool; Scotland; and Russia).[13] It is a list remarkable for the number of subscribers from the southwest interior—namely Kentucky, the Mississippi territory and Louisiana (the last, in particular, had as many subscribers as New York). The first volume of Rees's *Cyclopedia* would similarly evidence the reach of an increasingly national, Philadelphia/New York publishing axis, listing associated booksellers in Boston, Salem, Portsmouth, Portland, Baltimore, Washington, Georgetown, Alexandria, Fredricksburg, Richmond, Petersburg, Charleston, Savannah, Augusta, Pittsburgh, and Lexington, Kentucky.

11. The subscribers were listed, at the end of volume nine in the the first edition of the *Ornithology*, according to state or territory. Individual subscribers are listed in letters from Wilson to Daniel Miller written from Washington, 24 December 1808 (see *Letters*, 296); and to Samuel Bradford from Savannah, 8 March 1809 (*Letters*, 312). A list of the Virginia subscribers is included in Robert Cantwell, *Alexander Wilson: Naturalist and Pioneer* (Philadelphia, 1961), 277, 305.

12. After 1790, and prior to the coming of the railroad in the 1830s and 1840s, Philadelphia and New York established, and came to dominate, the book trade in the southern and western interior; see William Charvat, *Literary Publishing in America, 1790–1850* (Philadelphia, 1959): "Philadelphia publishers began to control the southern book-buying market, and New York the territory west of the Hudson, both cities sharing the trade of the Ohio valley." It was the leading publishers of Philadelphia and New York, rather than landlocked Boston, who, in the age of steamboats and canals, sought "the trade not only of the coast but of the interior"—and who, as Charvat argues, thereby "discovered (in a sense, established) the common denominator in the literary taste of the whole country." Charvat summarizes the transformation from diffuse to centralized and nationalized literary publishing with figures from fiction sales: "In the first decade of the nineteenth century, almost fifty per cent of our native fiction was published outside of New York, Boston, and Philadelphia—in the 1840s, only eight per cent" (pp. 23–24, 26).

13. Five of these subscriptions were for two or three copies; twenty-one were for institutions: city or college libraries, athenaeums, learned or medical societies, and, in the case of Pennsylvania, the Pennsylvania Legislatures (three copies) and the Pennsylvania Hospital (two copies).

Given its demographically elite but geographically extensive range, Wilson's *Ornithology* was, like Peale's museum, not only definitively Jeffersonian but also definitively federal—and federalizing. In the decades after the revolution, federalism was more fiction than fact. As Trish Loughran has argued, union was irreducibly virtual—a matter of fantasy and projection—in the founding, or federalist, period when the United States embraced a geography of communities so profoundly local and specific, so intensely diverse and disjunct, as to be actually unrepresentable.[14] In several of his travel letters, Wilson describes his reception on the road, not only by the "gentlemen" to whom he displays his sample volume but also by backcountry crowds who simply stare at the stranger in their streets, as if both the Philadelphia salesman and his bird book were "a bear or a mammoth," as he put it—a singular, portable museum display:

> I have laboured with the zeal of a knight errant in exhibiting this book of mine, wherever I went, traveling with it, like a beggar with his bantling, from town to town, and from one country to another. I have been loaded with praises, . . . shaken almost to pieces in stage coaches; have wandered among strangers, hearing the O's and Ah's, and telling the same story a thousand times over. . . .
>
> I spent the whole of this week traversing the street, from one particular house to another, till I believe, I became almost as well known as the public crier, or the clerk of the market, for I could frequently perceive gentlemen point me out to others as I passed with my book under my arm. . . .
> I reached Newark that day, having gratified the curiosity and feasted the eyes of a great number of people who repaid me with the most extravagant compliments, which I would have very willingly exchanged for a few simple *subscriptions*. . . . I spent nearly the whole of Saturday in Newark, where my book attracted as many starers as a bear or a mammoth would have done.[15]

14. Trish Loughran, "Virtual Nation: Problems of Representation in the Extended Republic," unpublished manuscript, 1997.
15. The first paragraph quotes a letter from Wilson to Daniel Miller, 26 October 1808 (*Letters*, 286); the latter two a letter from Wilson to a friend, Boston, 10 October 1808 (*Letters*, 276).

It was because of Wilson's journeys on behalf of *American Ornithology* that, some seventy years after Wilson's death, Henry Adams used his travel writings to help introduce his own nine-volume *History of the United States of America during the First Administration of Thomas Jefferson to the Second Administration of James Madison*. Specifically, Adams used Wilson's letters to support his contention that the sprawling Jeffersonian state was without adequate transportation, largely impenetrable, and—particularly in its far-flung interiors—an unenlightened, backward, even medieval place.[16] By Adams's account, Wilson himself was a cultivated but somewhat eccentric specimen—an "ornithologist, a Pennsylvania Scotchman, a confirmed grumbler, [and] a shrewd judge," but "the most thorough of American travellers."[17] What Adams fails to note, however, is that Wilson traveled not simply because he wished to but also because he had to. This compulsion derived in part from the literary market; not only did Wilson work for Bradford and Inskeep but, in imagining and designing his collection, Wilson also explicitly imitated other celebrated literary travelers: particularly his friend William Bartram, whom he met in 1803, and Lewis and Clark, who left for the Pacific in the same year. In fact, on the heels of Lewis and Clark's departure, Wilson had written to President Jefferson, hoping to be appointed to another government-sponsored western expedition up the Red River.[18] Neither this expedition nor Wilson's appointment would materialize, but Wilson continued to aspire to the status of the celebrated traveler. Conveniently for Wilson, Jefferson ordered the animal specimens collected by the Corps of Discovery to be sent to Peale's museum in Philadelphia: volume 3 of *American Ornithology*, therefore, exhibits some of the new western species gathered by Lewis and Clark (Lewis's woodpecker and Clark's crow) together with western species collected by Wilson on his own solitary expeditions (the Louisiana tanager, swamp sparrow, Savannah sparrow, water thrush, painted bunting, Mississippi kite, Tennessee warbler, Kentucky warbler, prairie warbler, Carolina parrot, blue-green and Nashville warblers, and so on).[19]

Both Wilson's nine-volume *Ornithology* and the forty-seven-volume *Cyclopedia* were eventually completed, but at a steep price. By 1818, the *Cyclopedia* was still unfinished and the project had cost $200,000, bankrupting the firm of Bradford and Inskeep. The remaining volumes of the *Cyclopedia* were completed by a syndicate of engravers in 1827, and the unsold volumes

16. Henry Adams, *History of the United States of America during the First Administration of Thomas Jefferson to the Second Administration of James Madison*, 9 vols. (New York, 1889–91).

17. Adams, *The United States in 1800* (Ithaca, N.Y., 1979), 13.

18. Wilson to Jefferson, 6 February 1806 (*Letters*, 249).

19. See Hunter, "Wilson's Principal Ornithological Travels" (*Letters*, 213–14).

dispensed by lottery.[20] Wilson, however, did not live to see even this equivocal outcome. Having assembled a remarkable seven volumes in just six years, zigzagging the eastern half of the continent in order to collect, describe, classify, and illustrate more than three hundred bird species, he died in 1813, quite literally of overwork—of "heart palpitations," chronic dysentery, and chronic poverty. When Peale painted *Exhumation of the Mastodon* in 1807, Wilson had just finished the first volume of his *Ornithology;* by the summer of 1813, he was in the midst of work on his eighth volume. He wrote to his father in Scotland: "Intense application to study has hurt me much. My 8th volume is now in the press. . . . One volume more will complete the whole."[21] After Wilson's death in 1813, the eighth and ninth volumes were completed by Charles-Lucien Bonaparte, Napoleon Bonaparte's expatriate nephew. The connection to Napoleon is more than coincidental. If the collections of Italian and German art brought back to Paris by Napoleon's armies were mementos of empire in Europe (and if Alexander Von Humboldt was only the latest voice of centuries of colonialist investment in South America), then Wilson's *Ornithology* is the peculiar expression of Jeffersonian empire in North America. Despite its rural or pastoral aspect as a field, the specificity of *American Ornithology*—its Jeffersonian/federalist stamp, so to speak—lies in its character as a performance of production, in its attempt to constitute not only American birds but also its author as both representative producer and celebrated employee of the federal state.

NATIONAL CHARACTERS/REPRESENTATIVE BIRDS

Despite the fact that John James Audubon saw his work as an imitation of, and in competition with, Wilson's *American Ornithology*, it is Audubon's *Birds of America* (London, 1827–38) that has been the more widely known and reprinted work. Audubon successfully represented himself as an adventuring frontiersman of science; his great birds have been viewed as a place where Romantic art meets natural science, where history painting meets book illustration, and even where artistic genius on a grand scale came into contact with the (supposedly mechanical truth of) particular detail, elevating these illustrations almost—but often, crucially, not quite—to the level of high art.

Like Wilson, Audubon also wrote an ornithological text, his *Ornithological Biography,* published in Edinburgh between 1831 and 1839, a five-volume series of short sketches in which—as in Wilson's text—self-advertising merges

20. Oberholtzer, *Literary History of Philadelphia,* 1:151; see also Tebbel, *Publishing in the United States,* 116, 174.
21. Wilson to David Wilson, Philadelphia, 6 July 1813 (*Letters,* 406).

fascinatingly with travel literature and anecdotes of natural history. The *Ornithological Biography* deserves more study than it has received from students of literature and culture, least of all because it elaborates and complicates the Romantic myth of Audubon the frontiersman, the ornithologist in buckskin. But this text is little read or known in comparison with his famous prints, especially those in *Birds of America*, which is viewed primarily as visual art, as a collection of beautiful pictures: it was produced in England for wealthy patrons on gigantic paper called "elephant folio," and then sold—separately from its ornithological/biographical text—in sets of five prints each. Overall, it has been as a great painter rather than writer of birds (birds painted, moreover, in the grand style by an artist who reportedly studied with the history painter Jacques Louis David) that Audubon has come to represent "American ornithology" and American nature in general in the imaginations of many North Americans: it is not for Wilson, after all, that the Audubon Society is named.

What this comparison should foreground is not that a hard-working Alexander Wilson has been treated unfairly by history in comparison to the more flamboyant Audubon, but that Wilson was, first of all (though not only), a writer and, more particularly, a poet: in Wilson's *Ornithology*, the linguistic text is as crucial to and as much a part of the ornithological project as are the engravings. Indeed, Wilson did not learn to draw until the age of forty, with help from William Bartram and Bartram's niece Anne. The *Ornithology* is a text written and illustrated by a man who saw himself as a literary traveler rather than as a visual artist or painter per se; it is in the tradition of Bartram and of Lewis and Clark (as well as British naturalists Gilbert White, Thomas Bewick, and Edward Latham)[22] that his *Ornithology* should be read—and viewed.

22. See W. B. Carnochan and Elizabeth Heckendorn Cook, *Gilbert White and the Natural History of Selborne* (Stanford, Calif., 1989). Writing in his journal in Liverpool, thirteen years after Wilson's death, Audubon described the British naturalist Thomas Bewick as "the Wilson of England" (entry of 4 August 1826, *The 1826 Journal of John James Audubon,* ed. Alice Ford Norman [Tulsa, Okla., 1967], 79). Thomas Bewick was a naturalist and engraver whose remarkably popular *History of Quadrupeds* had been published in 1790. This was followed by his two-volume *History of British Birds* (1797–1804), with its famous series of wood engravings. Audubon's comparison of Bewick to Wilson is significant because, like Wilson, Bewick had worked self-consciously to make his *History of British Birds* "a great national work." In his *Memoirs* Bewick writes of his respect not only for John Ray (a British taxonomist who "led the way to truth and to British Ornithology") but also for the ornithological works of Gilbert White, Thomas Pennant, and Edward Latham. Latham, whose system of classification would be adopted by Wilson, seemed particularly to Bewick to "have wound up the whole." At the same time however, Bewick "often lamented" that Latham's work "was not—by being embellished with correct figures—made a great national work, like the Count de Buffon's"; see *Memoir of Thomas Bewick, 1822–1833* (London, 1924), 131–32. Bewick sought to amend this defect with his own British ornithology. Numerous editions of Buffon's *Histoire Naturelle* (in forty-four volumes) of 1749–1804, as well as translations and adaptations from it (such as Oliver Goldsmith's eight-volume *History of the Earth and Animated Nature* [1774]) had initiated a growing popular interest in natural history that Bewick, Wilson, and later, Audubon would exploit and extend.

Certainly, Wilson's drawings can be viewed or framed alone; but ultimately his *American Ornithology* must be considered as an illustrated *text*, one that is much more profoundly mixed generically than anything Audubon would produce. Each essay was composed of fragments of, or forays into, various other genres: poems, anatomical analyses, theater reviews, travel anecdotes, sketches of curious characters, moral lessons, and transcriptions of bird "dialects" as well as letters and parts of letters from Jefferson, Bartram, Samuel Latham Mitchell, and other reputable American "gentlemen." Formally and stylistically, Wilson's *Ornithology* is an uneven rag-bag, resembling in its disjunctively woven way the architecture of the birds' nests he describes: "formed of a little loose hay, feathers of a guinea fowl, some wool, hanks of thread, hog's bristles, skeins of silk, pieces of cast snake skins, and dog's hair, . . . bits of rotten wood, fibres of dry stalks of weeds, pieces of paper, commonly newspapers, . . . the whole tightly sewed through and through with long horse hairs, interwoven with the silk of caterpillars and the inside lined with fine dry grass."[23]

A poet first, Wilson had published a volume of dialect and English-language *Poems* in Paisley (printed by J. Neilson in 1790), prior to his emigration to Philadelphia. Forced to flee Scotland in 1794 for participating in revolutionary Paineite politics, he worked variously in Pennsylvania and New Jersey as a weaver, peddler, and schoolteacher until meeting Bartram and deciding to make "a collection of all our finest birds." But, having begun his climb to authorship as a Scots dialect poet, Wilson included in *American Ornithology* not only natural history sketches but also long poems.

Wilson's longstanding ambition of being "distinguished in the literary world" had involved the assumption and abandonment of a series of authorial "personae." Wilson had first imitated Robert Burns, hoping to appeal to a British literary market which, increasingly, produced and distributed poems and poets as native productions.[24] In the United States, ten years later, it was literary travel rather than poetic "genius" which was crucial to the establishment and proliferation of post-war American literary markets wherever they intersected with European imperial expansion. Wilson therefore made himself a literary traveler. He first imitated Thomas Moore, in 1804. Moore had visited Philadelphia and made a journey to Niagara in 1803, describing both in verse;[25] Wilson himself, then, immediately made a journey to Niagara, in order

23. "Baltimore oriole," "Great-crested flycatcher," and White-eyed flycatcher," *American Ornithology*, Jardine, 1:20, 225, 306. References are given henceforward in the text.
24. For the best discussion of the contradictions that made Robert Burns and his poetry see Carol McGuirk, *Robert Burns and the Sentimental Era* (Athens, Ga., 1985), esp. 103–19.
25. Thomas Moore, letter from Niagara, 24 July 1804, quoted in Elizabeth McKinsey, *Niagara Falls: Icon of the American Sublime* (Cambridge, 1985), 41.

to write his own, much longer travel poem, "The Foresters: A Pedestrian Journey to the Falls of Niagara in 1804." As Elizabeth McKinsey describes the tradition of Niagara, "Thomas Moore and Alexander Wilson inaugurated what would be a torrent of poetry about the Falls. . . . soon there would be plays and tales . . . as well as countless magazine and occasional prose pieces that focused on the cataract."[26] After imitating Moore, Wilson next became a naturalist, emulating Alexander von Humboldt (who had also visited Philadelphia in 1804), Lewis and Clark, Zebulon Pike, and above all Bartram, the celebrated literary traveler to the countries of the Creeks, Choctaw, and Cherokee.[27]

As a traveler, Wilson was familiar with the newest technologies in water travel. An acquaintance of Robert Fulton, Wilson knew very well the benefits of steam.[28] But when he was finally able to make his own expedition west in 1810, he did not navigate the Ohio River in any of the new mechanical boats; instead he embarked from Pittsburgh in a one-man "batteau," which he self-consciously dubbed "The Ornithologist." There was a kind of precedence for travel by batteau in Wilson's journey to Niagara in 1804, when (according to his poetic account in "The Foresters") he and his companions traveled in a skiff dubbed "The Niagara."[29] However, Wilson claimed that the idea for traveling west in this solitary and Romantic way had been suggested to him in 1806 by the Philadelphia typefounder Mr. Ronaldson—who had traveled down the Ohio to St. Louis "with one companion in a small Batteau," where they purchased "a quantity" of the lead needed for making type.[30] At any rate, what is obvious here is that the boat inscribed with the name "The Ornithologist" was

26. McKinsey, *Niagara Falls*, 38.

27. Seeing "by the papers," Wilson wrote Bartram in 1806, "that Mr. Jefferson designs to employ persons to explore the shores of the Mississippi," Wilson's first plan was to travel from Wheeling to New Orleans as the "companion and assistant" of Bartram himself. In January of 1806 he asked Bartram what he thought "of laying our design before Mr. Jefferson with a view to procure his advice and recommendation to influential characters in the route. Could we procure his approbation and patronage, they would secure our success"; Wilson to Bartram, 27 January 1806 (*Letters*, 247). When Bartram, then seventy, refused, Wilson turned to Jefferson in the hope of being appointed to the projected government expedition "up the Red River." To Wilson's disappointment, Jefferson never responded to his letters, and the expedition was canceled. Wilson's letters to Bartram and Jefferson show that his efforts to be appointed to an expedition were tied to the task of gathering specimens for his *Ornithology;* Wilson to Jefferson, Kingsessing, 6 February 1806 (*Letters*, 249–50).

28. In 1807, Fulton had launched the first commercially successful steamboat in New York. Wilson himself called on Fulton in New York on October 2 of that year, in order to ask the engineer to verify "some particulars" in the *New Cyclopedia* entry on canals; see Hunter, "Life of Alexander Wilson" (*Letters*, 81–82); and Wilson to Samuel F. Bradford, New York, 2 October 1807 (*Letters*, 265–66).

29. Wilson, "The Foresters: A Poem Descriptive of a Pedestrian Journey to the Falls of Niagara in the Autumn of 1804." (Newtown, Pa., 1818), 63.

30. Wilson to Bartram, 27 January 1806 (*Letters*, 247).

part of Wilson's performance of production, a performance framed at the intersection of word and image, visual and verbal self-production. It was precisely this performative side of Wilson's ornithological pose that would later appeal to Audubon. In 1823, over a decade after Wilson's death, Audubon made his own pilgrimage from Philadelphia to Niagara Falls, after having visited Wilson's publishers, engraver, and editor in Philadelphia—where Audubon was "coldly received." At Niagara, Audubon recalled Wilson's long employment by Bradford and Inskeep when, signing the register at his inn at the Falls, he wrote after his name, "who, like Wilson, will ramble, but never like that great man die under the lash of a bookseller." Immediately after leaving Niagara for the southwest, Audubon again imitated Wilson's performance of ornithological self-production—by buying a skiff, or batteau, in Pittsburg to take him down the Ohio River.[31]

Wilson invariably attempted to represent himself as a visible object, framed to his own view, upon an imaginary international stage or display case. Like all works of natural history in the Linnaean mode, the productivist and representational artifacts of *American Ornithology* aspired to the Enlightenment status of the international, the global, and universal, and universally applicable. It was on this international stage or museum that Wilson wished to display himself as a new specimen: "the first" American ornithologist.

As an expression of extended republicanism, whereby federalism identified itself with "its" continent, *American Ornithology* pushes the trope of the "elevated" point of view to an ornithological extreme—in the form, so to speak, of the "bird's-eye view." The pun of the bird's-eye view may sound facetious as a way of describing this imperial elevation and extension of the national "roof," but, as Peale reflected in his gothic, taxidermical way (and as Constance Rourke recognized long ago in her studies of the tall tale), the punning conjunction of the outrageous and the actual marks much of the art/work of the early United States as forms of world-making.

Consider the king bird, or tyrant flycatcher (fig. 25). Wilson's king bird migrates every winter to the interior of Spanish South America, specifically Peru, where the "vast Maragnon" River flows to the Amazon. While celebrating the American king bird and his annual return from the "torrid regions" of South America to the more "temperate" North, Wilson produces an international moral geography: "Impelled by the powerful impressions of love, and the dictates of nature, [birds from South America] have winged their way to our more temperate and abundant regions, to rear their young in safety, and

31. See Joseph Kastner, *A Species of Eternity* (New York, 1977), 223.

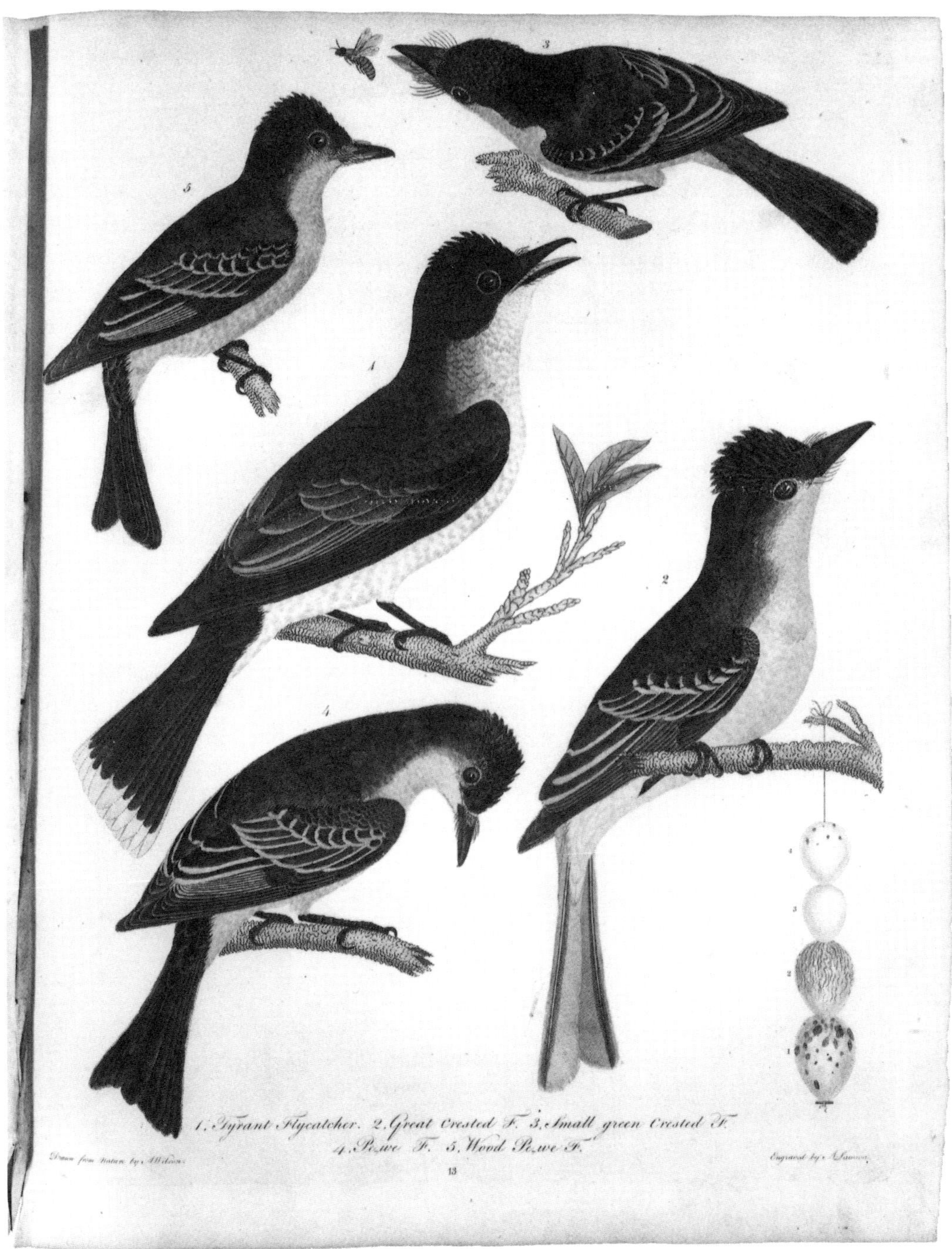

Figure 25. Tyrant flycatcher (no. 1), from Wilson, *American Ornithology*, volume 2 (1810). Huntington Library.

conduct a joyous progeny back to southern climes."[32] In *American Ornithology* the (Spanish-American) "South" is a place parents might return with their adolescent progeny for vacation or an educative tour. But the original nest-home is located in the North ("our more temperate regions"). As "torrid" places of (racial, moral, and vegetable) "color," South America and the Caribbean are exoticized objects of suspicion throughout *American Ornithology*, where color is "enjoyable" only so long as it can be incorporated and contained by the social taxonomies of the North. The scarlet tanager, for example, is "one of the gaudy foreigners (and perhaps the most showy) that regularly visits us from the torrid regions of the south": "While we consider him entitled to all the rights of hospitality, we may be permitted to examine a little into his character, and endeavour to discover whether he has anything else to recommend him, besides that of having a fine coat, and being a great traveler" (*Ornithology*, Jardine, 1:192). Dressed in a coat of "the richest scarlet, set off with the most jetty black," Wilson's tanager is a seducer or, perhaps, a lounger or dandy. The elevated and detached character of the traveler is as dangerous as his cultivation, and travel is desirable, finally, only so far as its fantastic height and breadth eventually take the traveler "home."

This is precisely what the king bird stands for, in a long narrative poem written by Wilson and embedded in his prose essay on the species. In this poem, a (single male) bird escapes from a colorful but "suffocating" South to return to the North. On the way, he surveys a map-like outline of topography which is also a moral geography:

> Far in the south, where vast Maragnon flows,
> And boundless forests unknown wilds enclose;
> Vine-tangled shores, and suffocating woods,
> Parched up with heat, or drowned with pouring floods;
> Where each extreme alternately prevails,
> And Nature sad their ravages bewails;
> Lo! high in the air, above those trackless wastes,
> With Spring's return the king bird hither hastes;
> Coasts the famed Gulf, and, from his height, explores
> Its thousand streams, its long indented shores,
> Its plains immense, wide opening on the day,

32. "On this hand he beholds one species just returned from the orange groves of Guiana and Surinam. . . . On that, a second, who has abandoned the shores of the Amazon or Orinoco, to trace his shallow streams and rivulets . . . while a brilliant and numerous family, in green, blue and glowing scarlet that but lately caught the eye of the sun-burnt savage of Brazil, amidst his native woods, now flutter through our orchards" ("Proposals," *Letters*, 269).

> Its lakes and isles, where feathered millions play:
> All tempt not him; till gazing from on high,
> Columbia's regions wide below him lie;
> There end his wanderings and his wish to roam,
> There lie his native woods, his fields, his *home;*
> Down circling, he descends, from azure heights,
> And on a full-blown sassafras alights.
>   (*Ornithology,* Jardine, 1:222)

Here "sassafras" iconically identifies (the northern) home, while the solitary king bird, foregrounded on his perch at the poem's end, is framed as a picture suitable for any Victorian parlor. In *American Ornithology,* it is the ultimately domestic nature of the king bird's homing impulse, bringing him north to his "nest," that marks him as a representative national species. Even the king bird's notorious aggression (which has earned him the name of "tyrant" flycatcher) is in the service of defending his family from plundering jays, thieving cuckoos, and even eagles. In aggressive self-defense, he becomes an ideal type of American Revolutionary virtue, fulfilling its principle of "common sense," whereby domestic attachments translate naturally and rightfully into aggression and even war:

> All danger over he hastens back elate,
> To guard his post, and feed his faithful mate . . .
> . . . Come now, ye cowards! . . .
> When the hour of danger and dismay
> Comes on your country, sneak in holes away,
> Shrink from the perils ye were bound to face,
> And leave those babes and country to disgrace;
> Come here (if such we have), ye dastard herd!
> And kneel in dust before this noble bird.
>   (*Ornithology,* Jardine, 1:222–23)

Again, the conclusion ("And kneel in dust before this noble bird") is iconic. In Wilson's bird collection, as in Peale's museum, it is the ability to frame visually and thereby "see"—and even to worship—nature that denotes elevation. And it is here that Wilson's *Ornithology* begins to fray at its Jeffersonian edges: as a purveyor of elevated views, Wilson puts himself sharply and repeatedly at odds with the American mechanic or agricultural producer whose labor grounds "Independence."

In its "museumizing" practices of classification and visual display, *American Ornithology* is preservationist in two senses of the word. On the one hand, it is a repository of specimens, a book version of the stuffed birds in Peale's museum. On the other hand, it is also, at many junctures, an environmentalist text arguing for the preservation and appreciation of birds, if not on the grounds of their entertaining characters and loving family life, then for their usefulness as killers of vermin.[33] But both of these forms of bird preservation require the bird-watching, spectatorial subject of *American Ornithology* to sever himself from the mechanics of production—the mechanics that ground Union—and book production. The "bird's-eye view" is thus diffused and established by Wilson's text at the cost of undermining its own federalist representational practices in an impossible effort to disjoin representative characters from the principle of production that raises them into view.

THE ANTIMECHANICAL TURN

The argument of Wilson's bird books turns again and again to a defense of nature's varied and distinctive characters over and against the unenlightened slaughter of American birds by farmers and mechanics. At such moments Wilson, a former handloom weaver, not only adopts a sentimental pose, mourning the murder of nature's creatures (because they live in "families"); he also assumes the point of view of a counterrevolutionary near-monarchist (or, given his Scottish background, a near-Jacobite):

> See where [the farmer] skulks! and takes his gloomy stand,
> The deep-charged musket hanging in his hand;
> And, gaunt for blood, he leans it on a rest,
> Prepared, and pointed at thy snow-white breast. . . .
> If e'er a family's griefs, a widow's woe,
> Have reached thy soul in mercy let him go!
> Kill not thy friend, who thy whole harvest shields.
> Some small return—some little right resign,
> And spare his life whose services are thine!
> —I plead in vain! Amid the bursting roar,
> The poor lost king bird welters in his gore!
> (*Ornithology,* Jardine, 1:223)

---

33. For a taxonomy of ornithological "discourses," see Kevin R. McNamara, "The Feathered Scribe: The Discourses of American Ornithology before 1800," *William and Mary Quarterly* 47 (1990): 210–34.

The poor lost king bird is displayed here for the education of uncultivated cultivators—frontier squatters and farmers, among other comparatively characterless sorts who supposedly cannot see the birds before them, and whose possession of the frontier is merely mechanical and, therefore, a-federal. In response, Wilson's "King Bird" essay sets forth a visual pedagogy that teaches the backcountry cultivator to elevate, see, and display—not only birds but also himself as a reader and viewer of nature.

Also at issue in Wilson's attempt to preserve American nature is the long-standing Jeffersonian argument with the French naturalist Buffon over the supposedly weak and degenerate character of American species. For Buffon, American woodpeckers were yet another example of the derivative (that is, the mechanical, atrophied, and generally more bug-like) nature of American nature.[34] Buffon compared American woodpeckers to "European peasants and mechanics," who lead lives of endless labor. "Constrained to drag out an insipid existence in boring the bark and hard fibres of trees," American woodpeckers in particular were, according to Buffon, mechanical producers of a low character, degraded by drudgery. But woodpeckers were not only maligned by Buffon; they were also regarded as "vermin" by American farmers, who shot them by the score: "This elegant bird is well known to our farmers and junior sportsmen who take every opportunity of destroying him for the supposed trespasses he commits on their Indian corn, or . . . for the trifle he will bring in market, . . . for the mere pleasure of destruction, and perhaps for the flavour of his flesh" (*Ornithology,* Jardine, 1:43–44). Defending the "innocent amusements" of the gold-winged woodpecker against both Buffon and the American farmer, Wilson asks his reader to take a closer look at the gold-winged woodpecker— and his looks (fig. 26): "The abject and degraded character," Wilson writes, "which the Count de Buffon . . . has drawn of the whole tribe of woodpeckers, belongs not to the elegant and sprightly bird now before us."

> *He* cannot be said to "lead a mean and gloomy life, without an intermission of labour," who usually feasts by the first peep of dawn, and spends the early and sweetest hours of morning on the highest peaks of the tallest trees . . . "
>
> Can it be said . . . that "the narrow circumference of a tree circumscribes his dull round of life," who as seasons and inclination inspire, roams from the frigid to the torrid zone, feasting on the abundance of various regions? Or is it a proof that "his

---

34. The history of this debate is detailed in Antonello Gerbi, *The Dispute of the New World: The History of a Polemic, 1750–1900,* trans. Jeremy Moyle (Pittsburgh, 1973).

> appetite is never softened by delicacy of taste," because he so
> often varies his bill of fare, occasionally preferring to animal
> food the rich milkiness of young Indian corn. . . ? Let the reader
> turn to the faithful representation of him given in the plate, and
> say whether his looks be "sad and melancholy." (*Ornithology*,
> Jardine, 1:50–51)

Wilson's imperative to look at the woodpecker's looks is performed by the plate to which the reader is directed (fig. 3): here, in a compositional arrangement common to European travel and magazine illustration, the position of the reader/viewer is represented within the plate by a smaller bird-spectator, who looks up admiringly at the large, handsome, and "happy" woodpecker above.[35] Displayed in a handsome engraving and described in vivid prose, this specimen is no mere "mechanic" but, rather, a citizen of the world, "roam[ing] from the frigid to the torrid zone" and capable of enjoying the varied diversions of "taste"—diversions that correspond with the "sportive, elegant" and (above all) varied pleasures of *American Ornithology* itself.[36] Inviting the reader/viewer to learn to see himself as a cultivated, and visually alert bird, Wilson's gold-winged woodpecker balances industry with a variety of respectable, leisure-time amusements. Capable of diversion, he is the subject and the object of those visual techniques of appreciation that *American Ornithology* aimed to disseminate: all bird hunters must become cultivated; all Anglo-Americans, particularly recent European immigrants and western emigrants, must become more elevated and enlightened, more visually literate. Again, the imperative condemns producers who forget the real (that is, visual, textual, representational, exhibitionary) principles of production whenever they go out to shoot gold-winged woodpeckers or king birds mechanically and impersonally: "the farmer . . . steals among the rows with his gun, bent on vengeance, and forgetful of the benevolent sentiment of the poet, that 'Just as wide of justice he must fall, / Who thinks all made for one, not one for all'" (*Ornithology*, Jardine, 1:46).

With this quotation, the poet suggests for a moment that the land is originally common, or "made for all." But this allusion to a propertyless economy is in the end just a sentimental memento. Wilson sympathizes finally with the economic values that justify shooting birds: "but most farmers," he concludes, "are, understandably, not much versed in poetry, and pretty well acquainted with value of corn, from the hard labour requisite in raising it" (*Ornithology*, Jardine, 1:46). Zigzagging, Wilson now allies himself with the Jeffersonian

---

35. McKinsey, *Niagara Falls*; see chaps. 1 and 2 for plates and a discussion of viewers within views.
36. Wilson, "Proposals for Publishing by Subscription . . . *American Ornithology*" (*Letters*, 269).

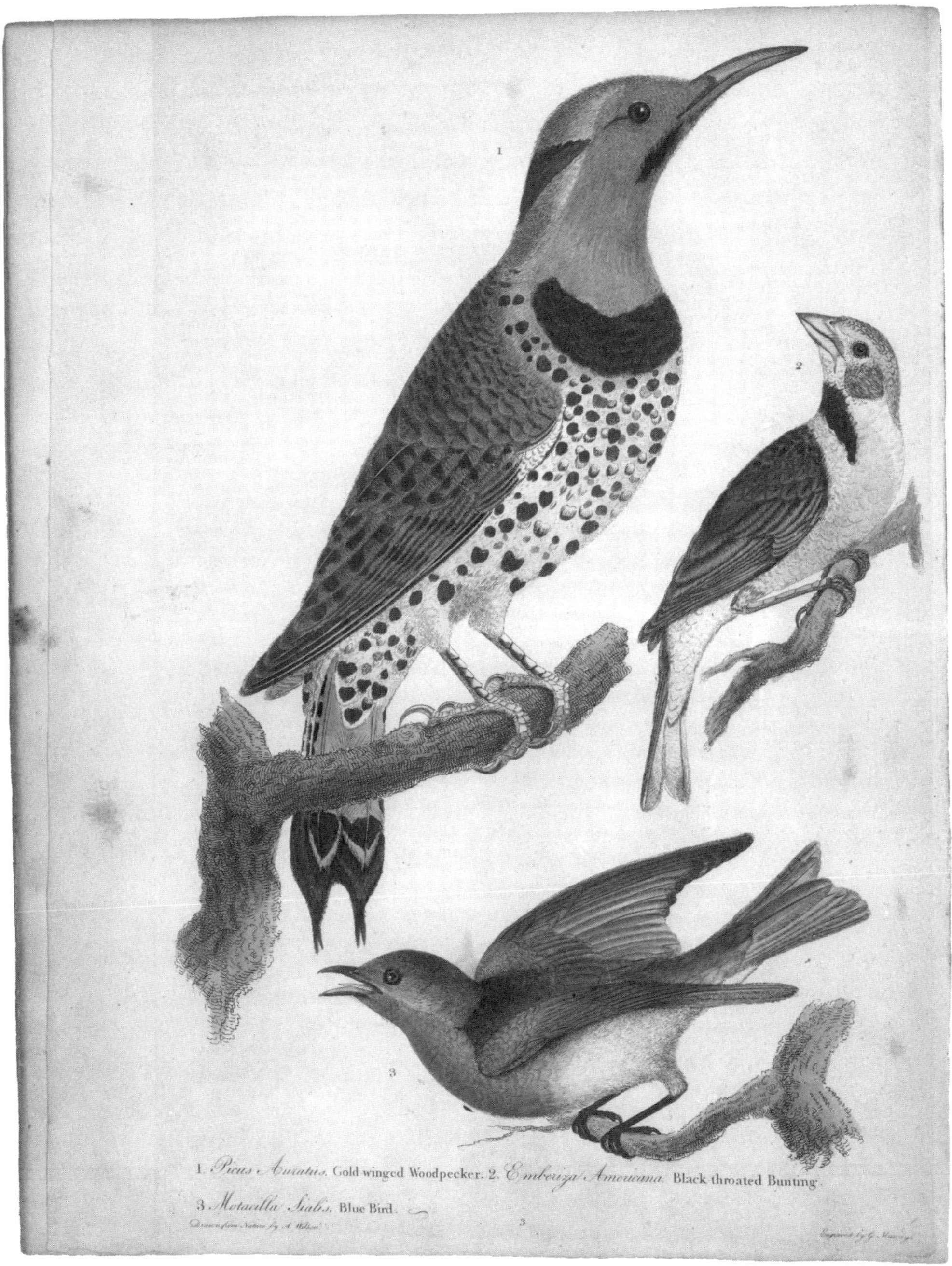

Figure 26. Gold-winged woodpecker (no. 1), from Wilson, *American Ornithology*, volume 1 (1808). Huntington Library.

husbandman and speaks understandingly of "the hard labour" requisite to "raising" corn, wheat—and the state. In the end, even though it repeatedly, and often agonizingly, tries to sever itself from its own ground in labor—as a work painstakingly, laboriously assembled—*American Ornithology* also ultimately reproduces, or raises again, a collective union via federal mechanics, (re)making itself and the world around it upon the same representational grounds it attacks. Finally, instead of a roof raised above a federal edifice or "mansion-house" (the central trope of Philadelphia's Grand Federal Procession of 1787), Wilson's *Ornithology* presents us with a less neatly centered and stable structure. As a federal and Jeffersonian collection, *American Ornithology* more closely resembles a huge tent covering acres of grass, a tent that is beginning to collapse here and there but that rises, baggily and doggedly, to any number of widely dispersed and fluttering—but pyramidal—points.

Elevation is figured, among other things, by the "big birds" in the collection—that is, those birds whose native characters were, in Wilson's view, the most representative. Each of the nine volumes contains between thirty and fifty species; but the essays written for each species vary radically in length, according to the representational significance, the peculiarities, or the familiarity of the species under description. Small birds, such as the Maryland yellow-throat, the small blue-gray flycatcher, and warblers generally, are given very cursory empirical descriptions, while others—such as the king bird, the gold-winged woodpecker, the mockingbird, the parrot, and the ivory-billed woodpecker—are occasions for extended socio-biological meditation upon that species' collective "character" and its implications for the character of American nature generally.

At the same time, *American Ornithology* is organized as a mere list of species, arranged in no special order, as they came into the "possession" of the ornithologist. Wilson stated explicitly that his *Ornithology* would not be organized according to the categories of systematic classification, which would have required him to open with "Acciperes" ("rapacious" birds) and, then "proceed regularly through the orders and genera" (that is, from Picae, "Pies," to Passerines, Columbe, etc.). Jefferson-like, the collector refuses to impose taxonomic order from above because, he says, "it is highly probable that numerous species at present unknown would come into our possession, . . . interrupting the regularity of the above arrangement."[37] Accordingly, it is the person and wanderings of the collector himself that are the ultimate point of

37. There is, however, an overall division into land birds (vols. 1–6) versus water birds (vols. 7–9), a division that deserves attention, but seems secondary; see Introduction, *Ornithology*, 1:vi.

reference—and Wilson appears repeatedly, therefore, in the text of *American Ornithology* as its productive author and traveler, a stand-in for the Maker or founder of all the species, who alone sustains, unifies, and justifies the whole sprawling assemblage.

"Shooting Ducks and Paroquets"

Wilson describes at length an episode in which he gratuitously shoots "paroquets," by which he meant Carolina parrots—the only species of parrot native to North America (fig. 27). Carolina parrots are now extinct, but when Wilson traveled through Louisiana, Kentucky, and Tennessee in 1810, he observed them as far north as Marietta, Ohio, even "in the month of February, along the banks of the Ohio, in a snow storm, flying about like pigeons, and in full cry" (*Ornithology*, Jardine, 1:378). Wilson adopted one of these parrots as a pet, named her Poll, and kept her in a cage although, upon being "remitted to 'durance vile' . . . it paid me in kind for the wound I had inflicted and for depriving it of liberty, by cutting and almost disabling several of my fingers with its sharp and powerful bill" (1:384). Poll served as a source of amusement and a traveling companion until the end of Wilson's journey. With Poll's cage in hand, Wilson attracted many curious viewers as he traveled through Louisiana. The bird proved especially useful when Wilson passed through Chickasaw and Choctaw country: "the Indians wherever I stopped to feed, collected around me, men, women, and children, laughing, and seemingly wonderfully amused with the novelty of my companion. Wherever I chanced to stop among these people, we soon became familiar with each other through the medium of Poll." Curious crowds of parrots likewise flocked "sympathetically" around Poll: "Numerous parties frequently alighted on the trees immediately above, keeping up a constant conversation with the prisoner" (1:385).

Despite Wilson's reiterated belief that the possession and exhibition of birds in cages would (like *American Ornithology* itself) have a softening and civilizing effect on their viewers, Wilson himself is a solitary shooter and gratuitous hunter of birds. In the midst of his anecdote about Poll, Wilson describes a sudden and unmotivated episode of slaughter. A flock of parrots had "come screaming through the woods in the morning, about an hour after sunrise, to drink the salt water. . . . When they alighted on the ground, it appeared . . . as if covered with a carpet of the richest green, orange, and yellow; they afterwards settled, in one body, on a neighboring tree, . . . covering

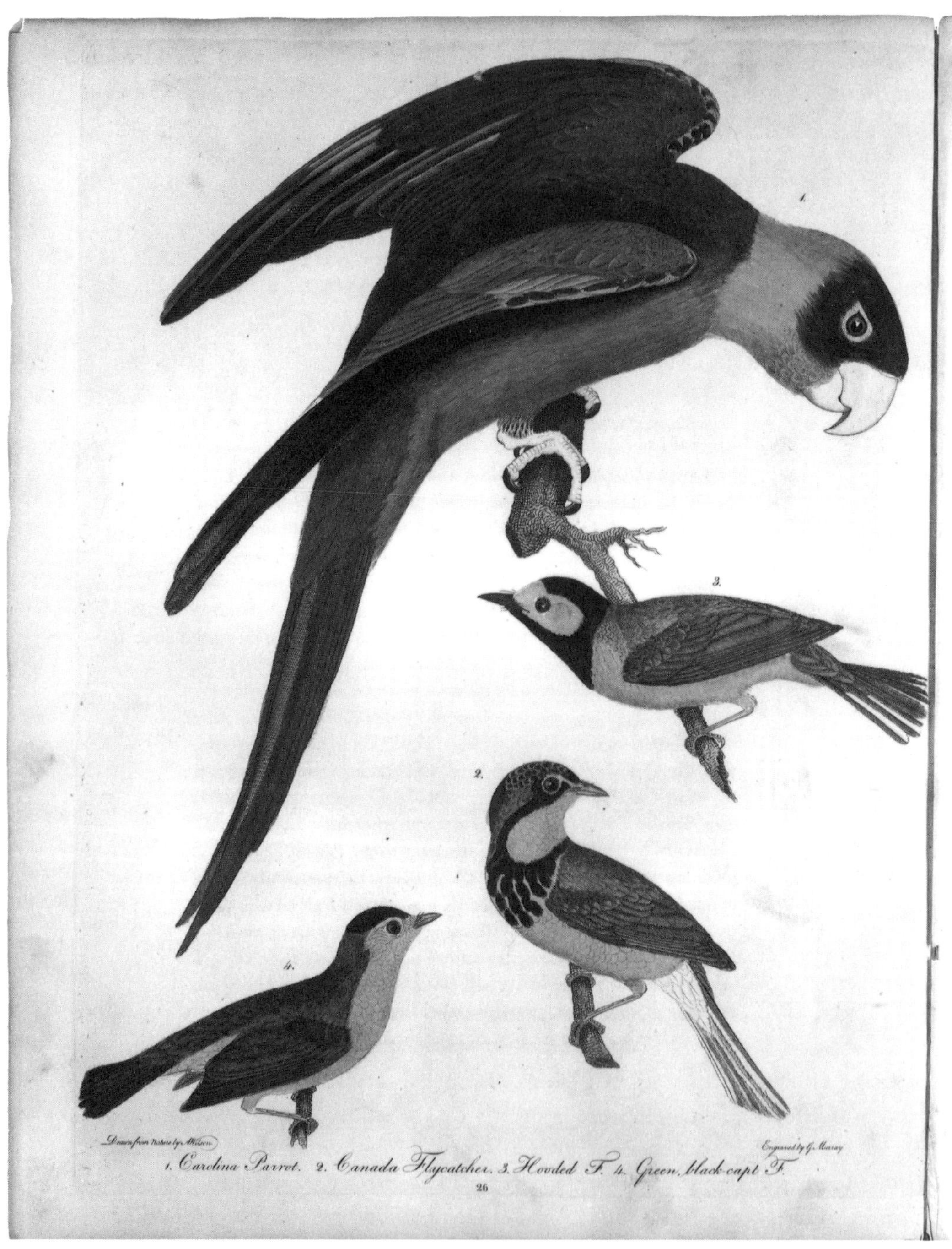

Figure 27. Carolina parrot (no. 1), from Wilson, *American Ornithology*, volume 3 (1811). Huntington Library.

almost every twig of it, and the sun, shining strongly on their gay and glossy plumage produced a very beautiful and splendid appearance." Then, without transition, Wilson begins to shoot:

> Having shot down a number, some of which were only wounded, the whole flock swept repeatedly around their prostrate companions, and again settled on a low tree, within twenty yards of the spot where I stood. At each successive discharge, though showers of them fell, yet the affection of the survivors seemed rather to increase; for after a few circuits around the place, they again alighted near me, looking down on their slaughtered companions with such manifest symptoms of sympathy and concern, as entirely disarmed me. (*Ornithology,* Jardine, 1:380)

American Ornithology presents such disjunctive moments of violence without commentary, in ways that seem split-off, half-conscious, and dissociated. Here, bird-killing is clearly not assimilable to the overriding ethic of the nest or of home production—where home production means the expansion of a United States domestic economy into brand-new interiors. But neither do such episodes of bird-killing stand in clear opposition to the expanding domestic scenery of the national nest. The often random shooting of "ducks and paroquets" (as well as blue jays, robins, rice birds, passenger pigeons, and others) in Wilson's *Ornithology* voices contradiction in a way that merely marks—without meaning much of anything by it—the sheer difference made by things that are not "things" (or commodities) and that cannot, therefore, be gathered to the national nest. The dissociation so evident here, between the act of slaughter and the moment of humanitarian disarmament, marks a place of breakdown in *American Ornithology.* It is precisely at this point—where Wilson can shoot down showers of birds and then be brought to heel by their "looks"—that the ornithological project of preservation, education, and cultivation is unable to assimilate, or absorb, the everyday violence of territorial appropriation, coerced work, and continual displacement to its own adoption of a cultivated tone.

Wilson continually instructs his audience in a laissez-faire ethic of benevolence: look, but don't touch; buy, read, or be visually absorbed, but don't shoot. At the same time, feathered death abounds. Wilson's pet Poll herself dies during the journey home to Philadelphia. Strangely gratuitous, this story also evades, without legitimating, *American Ornithology.* Upon reaching New Orleans, Wilson observed that Poll had become "restless and inconsolable" after the death of a companion parrot that had been caged with her. In response, Wilson performs an experiment:

> I placed a looking glass beside the place where she usually sat, and the instant she perceived her image, all her former fondness seemed to return, so that she could scarcely absent herself from it for a moment. . . . It was evident that she was completely deceived. Always when evening drew on . . . she laid her head close to that of the image in the glass, and began to doze with great composure and satisfaction. (*Ornithology,* Jardine, 1:386)

Immediately after Wilson's experiment with the mirror, however—in which Poll contentedly parroted the parrot—she plunges to her death: taking Poll with him on the ocean journey back to Philadelphia, Wilson was sleeping when "she wrought her way through the cage . . . flew overboard and perished in the Gulf of Mexico" (1:386). The experiment with the mirror can be chalked up to science; but Poll's death cannot, in fact, be absorbed by the ornithological project of cultivation—except where it foregrounds a representational framework that aims to let nothing escape its view. After playing her part as a representative and picturesque character, Wilson's parrot figures the particular losses (of particularity) constituted by *American Ornithology*—or what is left over, without significance, or without the federal state and its mirror of production.[38]

Throughout *American Ornithology,* cultivation implies a visual pedagogy according to which the ethic of elevation and the framework of visual display constitute a kind of cage. The parrot in her cage also allegorizes the emergence of authorship itself within a slowly bureaucratizing commercial culture where the objectification and isolation of its authors is masked by and as a structurally constitutive deception. Albeit in his silent, oblique, and split-off way—gratuitous and without irony—it is against this mirror, which frames him as the author and the subject of *American Ornithology,* that Wilson sometimes turned his gun (or his representational devices).

Paisley in the 1790s: The Worker Against Work

Between 1790 and 1795, Wilson had participated in and struggled directly against British industrial expansion in Scotland, and it was under the threat of imprisonment for sedition that Wilson emigrated to Philadelphia in 1794. His youthful crime was the writing and circulating of a dialect poem which had incited his fellow weavers to revolt. Like manufactory-era workers in nearby Glasgow, the weavers of Paisley had felt the brunt of social and economic changes introduced by foreign investment and the gradual mechanization of textile manufacturing at the end of the eighteenth century.[39] In the midst of the growing British reaction to the revolution in France and to growing anti-British sentiment in Scotland, Wilson had combined chronic poverty with weaving, book and cloth peddling, the writing of dialect poetry, and radical politics— specifically involvement in a Paineite Jacobin organization called "The Friends of Liberty and Reform."[40] Jacobin organizations in industrial west Scotland were able to connect the ideals and events of the French Revolution with Scots artisan/worker resistance to British industrial development, combining an artisan-based Scots nationalism with hostility to Britain's commercial and manufacturing expansion. It was not in the crafts and guilds of Edinburgh, but in the mills and manufactories of Glasgow and Paisley, that Wolfe Tone's "Society of United Irishmen" found counterpart Societies of United Scotsmen—united against a double "tyranny" of industry and empire.[41]

It was in this context that Wilson wrote the poem that occasioned his flight to Philadelphia. Titled "The Shark, or Lang Mills Detected," the offending poem described the fantasy beating of a Paisley mill owner, "Lang Willy Shark." Wilson had first been imprisoned in 1791 for another poem about a Paisley mill owner titled "The Hollander or Light-Weight," which had accused a Dutch silk manufacturer in Paisley of cheating his weavers, scrutinizing their work for holes, and weighing it for stolen thread. (The black market in goods woven from stolen thread was a traditional source of additional income for weavers in western Scotland.) The dialect poem for which Wilson was arrested in 1793, however, was even more egregious, as a seemingly direct incitement to violence:

38. See Jean Baudrillard, *The Mirror of Production,* trans. Mark Poster (St. Louis, 1975).
39. Norman Murray, *The Scottish Hand Loom Weavers, 1790–1850: A Social History* (Edinburgh, 1978), 40–50, 134, 168–172; and Thomas Crawford, "Political and Protest Songs in Eighteenth-Century Scotland II, Songs of the Left" *Scottish Studies* 14 (1970), pt. 2, 105–31. The handloom industry in west Scotland saw severe slumps (following cyclical booms) and its first strikes in 1787 and 1793 .
40. "Life" (*Letters,* 57–58).
41. See William Law Mathieson, *The Awakening of Scotland: A History from 1747–1797* (Glasgow, 1910); William Donaldson, *Popular Literature in Victorian Scotland: Language, Fiction and the Press* (Aberdeen, 1986), ix–35; and Brian Maidment, "Essayists and Artisans—The Making of Nineteenth-Century Self-taught Poets," *Literature and History* 9 (1983): 74–91.

> Whiles, in my sleep, methinks I see
>   Thee marching throught the city,
> And Hangman Jock, wi' girnan glee,
>   Proceeding to his duty.
> I see thy dismal phiz and back,
>   While Jock, his stroke to strengthen,
> Brings down his brows at every swack,
>   "I'll learn your frien' to lengthen,
>   Your mills the day."
>
> Poor wretch! in sic a dreadfu' hour
>   O' blude and dirt and hurry,
> What wad thy saftest luke or sour
>   Avail to stap their fury?
> Lang Mills, was rise around thy lugs
>   In mony a horrid volley;
> And thou be kicket to the dugs,
>   To think upo' thy folly
>   Ilk after day.[42]

It is poetry such as this, with its direct attack on the expansion of manu-facturing that is excluded from *American Ornithology*. After emigrating to Philadelphia, Wilson quickly came to regard his earlier participation in radical politics as a temporary aberration, a fumbling in the darkness. Apart from delivering a speech or two on behalf of candidate Jefferson, Wilson returned energetically and single-mindedly to his struggle to emerge from obscurity into the light of literary celebrity. In 1811, when Wilson's half-brother David arrived in Philadelphia from Scotland with copies of Wilson's early dialect poems, Wilson reportedly "threw them in the fire, unread, saying that if he had followed his worthy father's counsels they never should have seen the light."[43] He was then finishing the third volume of *American Ornithology* and had fin-ally begun to find a name for himself as the first American ornithologist—and as the friend of Bartram and Peale. Like Peale, Wilson repeatedly claimed at this time to have "long quitted the turbulent field of politics" for "the arts of peace" and the "peaceful, unassuming pages of AMERICAN ORNITHOLOGY": "Books on natural history, calculated to improve the taste, to enlarge the

---

42. *The Poems and Literary Prose of Alexander Wilson,* ed. Alexander B. Grosart, 2 vols. (Paisley, 1876), 2:61.
43. Grosart, "Memorial-introduction," *Poems,* 1:xxxviii.

understanding and better the heart, as they are friends to the whole human race, are generally welcomed by people of all parties."[44]

In spite of Wilson's wish to separate his sometimes violent, sometimes "licentious"—and supposedly failed—dialect poems of the early 1790s from his Jeffersonian *Ornithology*, the juxtaposition helps to illuminate the contradictions and disintegrations—the ragged edges or "holes" in—his "great work" of natural history. It seems safe to say, considering the episodes of bird-killing in Wilson's collection, that the rage and aggression provoked by the manufactory-era production and display of persons as things, and of authors as exemplary workers, found their own oblique articulation in *American Ornithology*. More obviously, however, if we rejoin the artisan-poet of the 1790s with the American ornithologist of 1807–14, a disjunction emerges between the federal collection and all that it failed to incorporate. By 1810, the cost of federal expansion included, among other things, a repression and forgetting (on the part of federalist art and technology) of the worker-governed wing of revolutionary social struggle as it had emerged in both Europe and the United States during the early years of the French Revolution. Among other things, what Wilson's great labor of American authorship explicitly refused to remember was the formation of increasingly class-conscious urban artisan organizations—the struggle and insight of manufactory-era artists and workers in both Europe and America.

Feathered Federalism

> . . . the sportsman need only take his stand near it, load, take aim, and fire; one flock succeeding another, with little interruption, almost the whole day: by this method, prodigious slaughter has been made among them with little fatigue. . . .
> In January, 1807, two young men, in one excursion after them, shot thirty dozen. (*Ornithology,* Jardine, 1:31–32)

Wilson's *Ornithology* does not only focus on individual specimens, however; it is also full of descriptions of huge flocks of birds: red-winged blackbirds, rice birds, grackles, passenger pigeons, robins, Carolina parrots, and others. In their overwhelming numbers—in flocks that sometimes take days to pass—birds in huge groups represent a kind of living surplus, a surplus that seems to escape, or fly free of, the relations of production. On the one

44. Wilson to [a Paisley friend] 3 November 1811 (*Letters*, 393); preface, *Ornithology*, vol. 3 (1811), v.

hand, then, the birds of America can be shot or harvested as food or as food for sale. But, on the other hand, insofar as birds, bird skins, or bird feathers are not of enormous market value, they also figure another, noneconomic kind of surplus, an elevated, aesthetic kind of surplus that seems, at first glance, to escape the mechanics of exchange: like the distinction between flowers grown for pleasure and crops grown for market, birds would seem to offer a form of enjoyment that could ratify national growth as something other than merely productive, commercial, or mechanical.[45] It is precisely as a living surplus, however, that American birds represented the possibility of other kinds of investment. Wilson repeatedly deploys the language and visual practices of "enjoyment" and "appreciation" as a way of recouping this second kind of surplus. In his discussion of the black-throated blue warbler, Wilson explicitly distinguished the ornithological "taste" for birds from the European taste for quadrupeds—and the virtues of the United States domestic economy from the British fur trade. Little is known of this warbler in Canada, Wilson writes, because "the feathered race is little known or attended to . . . on that part of the continent. . . . The habits of the bear, the deer, and beaver are much more interesting to those people,"

> and for a good substantial reason too, because more lucrative; and unless there should arrive an order from England for a cargo of skins of warblers and flycatchers, sufficient to make them an object worth speculation, we are likely to know as little of them hereafter. (*Ornithology,* Jardine, 1:261)

Unlike fur-bearing quadrupeds, but like the independent Jeffersonian republic, both the American "nest" and *American Ornithology* would seem to be aloof from the corruption of international markets. ("I candidly declare," Wilson writes, "that lucrative views have nothing to do in the business. . . . [A] wish to draw the attention of my fellow-citizens from the jarrings of politics, to the contemplation of the grandeur, harmony, and wonderful variety of nature . . . are my principle . . . motives" [Introduction, *Ornithology,* vol. 1]). In *American Ornithology,* bird-watching is primarily a "rural sport," a leisure-time pursuit that cultivates surplus as visual enjoyment or aesthetic appreciation. At the same time, however, the visual framework of *American Ornithology* articulates a landscape of violence, contradiction, and rupture, particularly at those moments when it represents Wilson slaughtering birds gratuitously, or unable to contain the diversity he

---

45. Harriet Ritvo's work on the semiotics of quadrapeds in Victorian England reveals an important contrast with the centrality of ornithology to the iconology of Jeffersonian empire; see Ritvo, *The Animal Estate: The English and Other Creatures in the Victorian Age* (Cambridge, Mass., 1985).

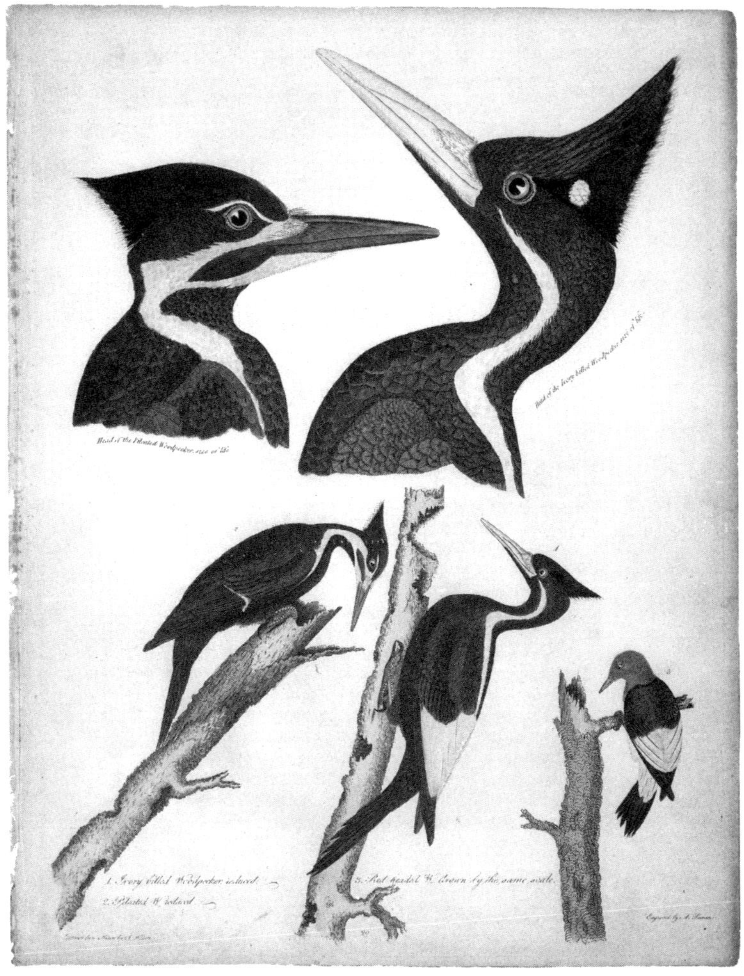

Figure 28. Ivory-billed woodpecker (no. 1), from Wilson, *American Ornithology,* volume 4 (1811). Huntington Library.

pursues. At such sites of rupture, *American Ornithology* simply fails to ratify the representational mechanics of federalism and one finds that birds do, in fact, escape the net of use-and-enjoyment. At the violent edges of *American Ornithology,* the visual and verbal mechanics of federalism cannot seamlessly contain the mammoth state they project.

Some of the most graphic episodes of breakdown in Wilson's *Ornithology* correspond with entries that describe birds inhabiting the Southeast and Southwest, the land of the Cherokee, Muskogee, Choctaw, and Chickasaw. In the "Ivory-Billed Woodpecker" (fig. 28), for example, the entry that opens volume 4, Wilson compares his specimen to both a baby and an "Indian."

Adorned in miniature with ivory (the commodity that helped open Africa and India to European capital), the ivory-billed woodpecker is pronounced doomed to "extinction," and Wilson analogizes the species to Native Americans who reject adoption, assimilation, and the plough. A "royal hunter . . . the king or chief of his tribe, . . . ornamented with carmine crest and polished ivory," the bird is ornamented like "the southern tribes" (that is, Creek, Seminole, Choctaw, Chickasaw) who, Wilson reports, ornament themselves, in turn, with the head and bill, skin or feathers of birds: "Thus I have seen a coat made of the skins, heads and claws of the raven, and caps stuck round with heads of butcher birds, hawks and eagles."[46] In the swamps of North Carolina, Wilson captures a member of this species, and decides to keep it alive to use as a model for his drawing. The bird is inconsolable, however, and its screams in captivity "exactly resemble the violent crying of a young child." Carrying his miniature captive "under cover," Wilson soon arrives at Wilmington, North Carolina. As he enters town, the bird's "affecting cries surprised everyone within hearing, particularly the females, who hurried to the doors and windows with looks of alarm and anxiety" (*Ornithology*, Jardine, 2:13). When Wilson arrives at his "hotel," the landlord, "alarmed at what he heard, asked whether he could furnish me with accommodations for myself and my baby." The American ornithologist quickly relieves everyone's anxiety: he produces his "baby," withdrawing the bird "from under the cover," while "a general laugh [takes] place." But the ivory-billed woodpecker himself comes to a less-than-humorous end. Left in Wilson's hotel room, the woodpecker wreaks havoc, breaking a fifteen-inch hole through the wall, covering the bed with plaster and nearly escaping. Wilson tries tying the bird to a mahogany table—upon which it likewise "wreaked his whole vengeance," nearly destroying it. While he was taking drawings of the woodpecker, Wilson reports, the bird "cut me severely in several places" and "displayed such a noble and unconquerable spirit that I was frequently tempted to restore him to his native woods. . . . He lived with me nearly three days, but refused all sustenance, and I witnessed his death with regret" (*Ornithology*, Jardine, 2:14). The head and body of Wilson's woodpecker remain, however, captured by Wilson's drawing; but, more centrally, as both feathered hunter and adopted infant, the ivory-billed woodpecker lives on as an artifact of the national family in its memories, jokes, and stories.

The principle of assimilation is familiar to historians of Jefferson and his administration: many scholars have noted Jefferson's rejection of military force

---

46. The "southern Indians" are said to believe that feathered ornaments, used as "amulets or charms, . . . confer on the wearer all the virtues or excellences of those birds" (*Ornithology*, Jardine, 2:14).

in favor of a policy that aimed to convert North American Indians to commercial farming via education, example, and rituals of adoption.[47] Jefferson himself advocated the peaceable assimilation (through cultivation) not only of Native American people but also of all "unproductive" tribes, including recent European immigrants and Anglo-American "squatters."[48] In attempts to convert a-federal people to federal land-use policy, Jefferson and his agents theoretically preferred a politics of example and performance to the overt violence of forced removal.[49] In both the "Ivory-Billed Woodpecker" and the "Carolina Parrot," the affecting spectacle of a captive bird is made to serve as a mediator of contact, a softening agent of Union.

In 1799, Peale had summarized the affective philosophy of his museum in a series of lectures delivered at the University of Pennsylvania. The 1790s had been a period of counterrevolution and reaction in Philadelphia (as well as Paris). Peale is concerned, therefore, to argue that his museum is a deeply moral rather than libertine institution, a virtuous rather than prurient display of mother nature's "parts." From birds to bears, Peale declares, social harmony characterizes nature's nations. Contrasting the American cuckoo to the European cuckoo—a notorious libertine and the "prototype of cuckoldry"—he argues that the secrets of American nature support a peace-loving domestic order:

> Here I feel gratification, nay, pleasure, that I am able to show
> this nest and Eggs belonging to these American Cuckoos. They
> build their own nest; they foster their young; they chaunt their
> soft notes to sweeten the care of Incubation—and I am proud
> to believe that they are faithful and constant to each other.[50]

Although this example was presented tongue-in-cheek, Peale was completely serious about the social virtues that he saw displayed by nature—in miniature.

47. See Joel W. Martin, *Sacred Revolt: The Muskogees' Struggle for a New World* (Boston, 1991), 114–86; and Bernard Sheehan, *Seeds of Extinction: Jeffersonian Philanthropy and the American Indian* (Chapel Hill, N.C., 1973). The U.S. Creek Treaty of 1790, signed in New York, is a useful example of the failure of models of national representativeness when applied to an Amerindian alliance groups; see John Walton Caughey, *McGillivray of the Creeks* (Norman, Okla., 1938; 1959). For a wider discussion, which considers the possibility that native American "examples of representative democratic social organization and leadership were formatively influential" and productive of (rather than merely appropriated or erased by) federalizing models of democratic-republicanism, confederation, and statesmanship see Donald A. Grinde Jr. and Bruce E. Johansen, *Exemplar of Liberty, Native America and the Evolution of Democracy* (Los Angeles, 1991), 1–59.

48. See Martin, *Sacred Revolt,* 87–116.

49. See, for example, the policies and rituals that shaped the encounters of Lewis and Clark's Corps of Discovery with the Oto and Yankton Sioux as summarized by Stephen E. Ambrose, *Undaunted Courage: Merriwether Lewis, Thomas Jefferson, and the Opening of the American West* (New York, 1996), 155–64.

50. Charles Willson Peale, "Lecture on Natural History," no. 18 (1799–1800), 12; quoted in Sellers, *Mr. Peale's Museum,* 109.

In another such lecture, delivered at the University of Pennsylvania, Peale imagines a world of animal nature penetrated at every level by "constancy and parental care":

> —Our traders to the Faulkland and other Isles . . . tell us that Sea-Lyons, Sea-Wolves, and other such animals, are often found so numerous, as to cover the shores; so thickly are they inhabited, that man with a short bludgeon, may kill hundreds of them before breakfast!—yet among this immense number of creatures a perfect harmony prevails!
> —Suppose we descend, and view the smaller animals,—here, myriads of insects present themselves to our view . . . behold among them also, a perfect harmony . . .
> And if we study the manners of such animals in general, we shall find amongst them, most excellent models of friendship, constancy, parental care and other social virtues.[51]

In this passage, Peale juxtaposes a violent foreground—the "man with a short bludgeon"—with the miniature world of birds and other animals. While it was his lifework to depict the lives of animals as "models" of human happiness, Peale himself killed most of the birds on display in his museum. His letters and diaries document countless shooting expeditions to obtain bird specimens, both for his museum and for exchange with institutions in Europe.[52] The museum mythology of Jeffersonian political economy attempted to fuse a domestic ideology with territorial appropriation and the violence of republican representation (as in the harvesting of animals for skins, feathers, or specimens for museum display). By virtue of the seamlessness with which it fused what Peale called "Art and Nature,"—or, one might say, interior and exterior habitats—the microcosm of the family and, more specifically, the family farm itself provided both Jefferson and Peale with a national icon that could seemingly regenerate itself no matter how often it was attacked and destroyed. As an American Noah's Ark, a self-sufficient domestic sphere was survivable, and would always come out on top, or so it seemed. As the locus of natural reproduction, the "nest" was an endlessly regenerative icon: "hundreds" of little worlds might be demolished "before

---

51. Peale, "Discourse Introductory to a Course of Lectures on the Science of Nature," 8 November 1800, delivered in the Hall of the University of Pennsylvania, and printed in Philadelphia by Zachariah Poulson Jr. (quotation on p. 10).
52. "Peale reflect[ed] on the waste he had made of the feathered tribe in order to furnish his Museum"; Charles Willson Peale, typescript autobiography, *The Collected Papers of Charles Willson Peale and His Family*, ed. Lillian B. Miller, Kraus Microform (Millwood, N.Y., 1980), Series II–C, 202, 205, 244.

breakfast" and still stand—as they did in the bird cases of Peale's Museum—as household shrines to national peace and harmony.

In the introduction to his *Ornithology*, Wilson compares himself to a schoolboy, home for the holidays, and his *Ornithology* to a "bouquet" of wildflowers, which he brings home to his mother: "Look my dear 'ma, what beautiful flowers I have found growing on our place. Why all the woods are full of them! red, orange, blue and 'most every color. O I can gather you a whole parcel . . . of them, much handsomer than these, all growing in our own woods!" He goes running out, "on the wings of ecstasy," to gather more. Wilson writes:

> Should my country receive with the same gracious indulgence the specimens which I here humbly present her, should she express a desire for me *to go and bring her more,* the highest wishes of my ambition will be gratified; for, in the language of my little friend, *our whole woods are full of them!* and I can collect hundreds more, *much handsomer than these. —A. W.* (Preface, *Ornithology,* 1:i–ii)

Here, where the labor of production is both "natural and spontaneous," rooted in the child's "love" for his mother, it ends in possession. Where the little boy's world of nature is "our place" and "our very own woods," ownership is presumed as a kind of natural title to animals and flowers. In the figure of the American mother, both Home and Nation are, in other words, everywhere—wherever specimens of natural history are to be gathered. And it is here that *American Ornithology* maps and records, projects and produces, a national "interior" that is, simultaneously, a continental and an emotional or feeling state. In *American Ornithology*, "the nest" went west Jeffersonian style: the nation is the continent—its incomprehensibly diverse "interior" is "our place"—while the imperial expansion, the sprawling migrations, and the extensions of American manufacturing are all marked as ultimately "domestic," homebound, or made in America.

The failures of *American Ornithology* to "fly"—the unwieldiness of the sprawling collection and its inability, in the end, to "raise" its maker into the light (of fame) he craved—are summarized by an anecdote from one of Wilson's travel letters from Kentucky. In 1810 Wilson went bird hunting at the Big Bone Lick, where an "enormous congregation" of mammoth skeletons were known to be sunk or buried beneath the mud. Having propelled his batteau into the Big-Bone Creek, Wilson "amuses" himself for a time "shooting ducks and paroquets" at "that great antediluvian rendezvous of the American elephants." While pursuing a bird across the Lick, Wilson feels himself beginning to sink:

> In pursuing a wounded duck across this quagmire, I had nearly deposited my carcass among the grand congregation of mammoths below, having sunk up to the middle, and had a hard struggling to get out.[53]

This description aptly represents the kind of hybrid, border-life Wilson led as an immigrant poet in pursuit of literary independence. It also encodes the ambivalent, ultimately tenuous position Wilson sustained on the expanding ground of the federal state he served. Caught curiously between union and specificity, between resistance to and deployment of the representational practices that produced him together with his collection, Wilson is caught, so to speak, between the mammoth and the bird. Contrary to the situation of Peale, who was similarly positioned between mammoth and birds (the one and the many) in *The Artist and His Museum* (see fig. 1 in David Brigham's essay in this volume), Wilson finds the ground (of self-assembly) giving way beneath his feet. As an objectified and buoyant "work," manufactured by a market that assembled, displayed, and sent him forth as an exemplary producer, this particular American manufacturer feels himself drawn into a swamp of "grand" (republican and federalist) carcasses—the remains of the great works and workers that have gone before him.

University of Iowa

---

53. Wilson to Alexander Lawson, Lexington, 4 April 1810 (*Letters*, 335).

By the Book: Audubon and the Tradition of Ornithological Illustration

Linda Dugan Partridge

All too familiar is the story of John James Audubon as a self-taught woodsman who followed only nature and his genius in painting American birds. He was indebted to the Old World, according to the tale, only for "discovering" his talent in 1826, when he sought a British publisher for *Birds of America*. Because the legend is endearing, it endures.

Aubudon himself was highly effective in hand-tailoring his image for posterity. At the age of forty-five, he recounted his youthful frustrations in learning to draw birds:

> I turned to my father, and made known to him my disappointment and anxiety. He produced a book of Illustrations. A new life ran in my veins. I turned over the leaves with avidity; and although what I saw was not what I longed for, it gave me a desire to copy nature. To Nature I went, and tried to imitate her.[1]

Elsewhere, Audubon wrote of his effort to please his viewer "by adopting a different course of representation from the mere profile-like cut figures given usually in works of that kind."[2] His contention was that, once having seen the old style, he soon rejected it and responded exclusively to the living bird.

That, however, was simply not the case. Prompted by his father's interest, he was aware of illustrated ornithologies from his early youth. As an adult he had access to libraries with ornithological sources through his contacts with scientists in France in 1805, New York City in 1806, and Cincinnati in 1820; and his periodic visits to Philadelphia culminated in his association with members of the Academy of Natural Sciences in 1824.

---

1. Audubon, "Introductory Address," in *Ornithological Biography* (Edinburgh, 1831–39), 1:vii.
2. Audubon, *My Style of Drawing Birds* (1828; reprint, New York, 1979), 21.

✎ 97

During the years he lived in America, his friends and in-laws included wealthy doctors and planters in the Ohio and Mississippi Valleys who held collections of natural history books. Between 1805 and 1826 Audubon consulted his own copies of William Turton's translation of Linnaeus and of the *American Ornithology* by Alexander Wilson and Charles-Lucien Bonaparte. He later amassed a costly library on birds.

The sheer size of Audubon's body of deceptively simple bird drawings has in fact confounded most attempts to analyze his oeuvre systematically or to track his development over time.[3] Even when the existence of his sources is acknowledged, it is easy to underestimate not only their range and complexity but also their continued value for Audubon. Instead of examining the evidence that the drawings provide, scholars have typically worked deductively from Audubon's known biography to posit which events influenced his art. For example, some have considered 1806–7 a quiescent period, when Audubon was chiefly occupied with courting his wife-to-be and with acquiring a trade. Yet this was a period of great stylistic experimentation, in which Audubon was actively engaged with contemporary science and illustration.

To chart the drawings' inscriptions, subjects, and style, I constructed a database analyzing over 190 drawings of the period 1805–26. The results show that his inscriptions referenced published sources regularly through 1812 and even as late as 1822. From 1805 to 1807 he noted the work of at least nine ornithologists, English, French, and American. Many of Audubon's identifications were based on the work of the French naturalist Georges-Louis Leclerc, Comte de Buffon, whose *Histoire naturelle, générale et particulière* (1749–88), with ten of its forty-four volumes devoted to birds, had been the single most popular work on natural history. Only two contemporary French titles on any subject sold better.[4] One of the many editions of Buffon was very likely a part

3. For instance, see Marshall B. Davidson, introduction, *The Original Water-Color Paintings by John James Audubon for* The Birds of America, 2 vols. (New York, 1966), 1:xxiii; Gary A. Reynolds, *John James Audubon and His Sons* (New York, 1982), 25; Edward H. Dwight, *Audubon: Watercolors and Drawings* (Utica, N.Y., 1965), 15, 9; Dwight, "Unpublished Audubon Originals," *Antiques Magazine* 87 (April 1965) 454–55; Dwight, "Audubon in Kentucky," *Antiques Magazine* 105 (April 1974), 853; Michael Harwood, "A Watershed for Ornithology," in *The Bicentennial of John James Audubon*, ed. Alton A. Lindsey (Bloomington, Ind., 1985), 43; Alice Ford, *John James Audubon: A Biography* (New York, 1988), 134; and Stanley Clisby Arthur, *Audubon: An Intimate Life of the American Woodsman* (New Orleans, 1937), 198. Only recently has his work been approached systematically, with the important 1993–94 traveling exhibition of Audubon's drawings held by the New-York Historical Society. See the essays by Theodore E. Stebbins Jr., Annette Blaugrund, Amy R. W. Meyers, and Reba Fishman Snyder, in the exhibit catalogue, edited by Blaugrund and Stebbins, *John James Audubon: The Watercolors for* The Birds of America (New York, 1993).

4. Paul L. Farber, "Buffon and the Concept of Species," *Journal of the History of Biology* 5 (autumn 1972): 282. Farber cites a study by Daniel Mornet, "Les Enseignements des bibliothèques privées (1750–1780)," *Revue d'histoire littéraire de la France* 17 (1910): 460.

of Audubon's father's collection in France; and if Audubon did not have his own copies, he at least consulted them in American libraries. By 1820, when he worked for the Western Museum in Cincinnati, the public library there owned a two-volume abridged set of Buffon's natural history and an index to his ornithology.[5]

While he claimed to be rejecting past tradition, Audubon constructed his own life's work, and the image he presented of his role, on the Baconian premise that his contributions would surpass those of others in the *progressive* discovery of a knowable truth about God's Creation—that is, the underlying model was that of a continuum. Were we to construct a simple chronology of science, and to fit Audubon within it, we could easily fall into this sort of formulation ourselves, finding a continuous development from the Linnaean emphasis on a static catalogue of immutable species toward a dynamic concept of life processes. The temptation to privilege and sequence the events that apparently led to our own technologies and theories seems doubly compelling when applied to artifacts produced in the era of positivism.

Historian Paul L. Farber proposes that instead of a single line of development there were at least two modes of natural history research germane to ornithology, pursued simultaneously in the eighteenth century.[6] One group of naturalists, including the Englishman John Latham and the Frenchman Mathurin Jacques Brisson, did indeed devote themselves largely to systems of classification. Another group, contemporary with the first, followed Buffon in studying the habits of the living bird. Buffon conceived of the Great Chain of Being not, as he put it, as

> a simple thread which extends only in length; it is a large woof
> or rather a bundle which from interval to interval sends out
> side branches to join with the bundles of another category.[7]

---

5. *Systematic Catalogue of Books Belonging to the Circulating Library Society of Cincinnati* (Cincinnati, 1816), n.p.

6. Paul L. Farber, *The Emergence of Ornithology as a Scientific Discipline, 1760–1850* (Dordrecht, Holland, 1982), xx–xxi. This criticism is frequently leveled at histories of science based on what scientists have claimed about their own discoveries. Michel Foucault's notion of the Classical age as characterized by correspondence between linguistic representation and the Linnaean mode of ordering natural history is especially useful in considering much early ornithological illustration. However, although Foucault is undoubtedly correct that Buffon organized his information on a Linnaean matrix, he fails to acknowledge Buffon's insistence on the primacy of the individual over artificial groupings and his interest in animal behavior. Because it is precisely these concerns with classification and the individual that influence Audubon (whose work exhibits features of both of Foucault's epistemes) and that in part characterize his images of individuals in dramatic conflict with random outcomes, Farber's model is the more helpful one for an analysis of this artist's work. See Foucault, *The Order of Things: An Archaeology of the Human Sciences* (New York, 1970), 135.

7. From Buffon, *Histoire naturelle des oiseaux*, 1:394, as quoted in Farber, *Emergence of Ornithology*, 24.

Buffon's dynamic model posited that some species, under some conditions, have capacity for change.[8] Farber's view is a useful one for studying Audubon; for the more eighteenth- and nineteenth-century sources one studies, the more evident it becomes that there is indeed no monolithic, progressive sequence of ornithological illustration to which Audubon brought a happy climax—that the term "traditional" itself is misleading.

In the following discussion, I suggest that Audubon neither broke from "tradition" nor carried it forward in progressive fashion; nonetheless, his achievement is intelligible only in terms of the scientific and aesthetic context of the late eighteenth and early nineteenth centuries. I do not tackle here the ambitious task of analyzing European ornithological imagery, but I wish to set out a framework of visual source material, and a horizon of interpretation, against which Audubon's entire oeuvre can be viewed.[9] Traditional ornithological illustration was a resource rich enough for Audubon to mine repeatedly and for varied ends throughout his career. Although most illustrations, taken separately, do not embody discrete ideologies, I will indicate how specific illustrators influenced Audubon and how pictorial conventions influenced his depictions, both generally and of certain bird types.[10] Traditional ornithologies even influenced the physical presentation of his drawings.

8.   Ibid., 126–27.

9.   Elsewhere I have analyzed the iconography of selected Audubon drawings for their scientific and theoretical bases, as well as the connections between illustration and text ("From Nature: John James Audubon's Drawings and Watercolors, 1805–1826" [Ph.D. diss., University of Delaware, 1992]; "Domestic Violence: Scientific Themes in Audubon's Rattlesnake Attacked by Mockingbirds," paper presented at the American Studies Association conference, November 1992; "A True Knowledge of the Birds: Audubon on the Frontier of Science and Art," paper presented at symposium on Audubon, St. Louis Mercantile Library, November 1994; and "Tough Chicks: Rethinking Relationships in Audubon's Feathered Tribes," paper presented at the New-York Historical Society, March 1997. See also Ann Shelby Blum, introduction, *Picturing Nature: American Nineteenth-Century Zoological Illustration* (Princeton, N.J., 1993).

10. Very little scholarship to date has addressed the topic of ideologies embedded within the pictorial conventions of the old ornithologies; for one of the few such discussions, see Blum, *Picturing Nature*, although it is is concerned with American rather than European documents. Illustrations can be conceptually loaded with Deism on the one hand, or natural theology on the other. Many systematists followed the early Linnaean doctrine of fixed species, or parallel concepts of a universally "ideal" nature embodied in "typical" forms. Others, Buffon included, believed some species to be mutable, and therefore subject to external pressures of, or interaction with, environment or climate. Similarly, the supposed Great Chain of Being was for some an endless, prioritized chain of species set in place all at the initial moment of Creation; others advocated a notion more like Buffon's, of bundles from various categories joining at intervals (see text, above, and Farber, *Emergence of Ornithology*, 24). Some authors chose to classify species' external morphologies; others described life habits and behavior.

The economics of book production also shaped the final appearance of illustrations. Book production—whether a capitalist or government-sponsored enterprise—was intended in most cases to attract wealthy patrons with the appearance of opulence. Such luxury dictated not only the physical appointments of the book but also the presentation of the bird as a consumable good. This could be accomplished through representation of sensuous textures, sinuous contours, or feather coloration. It ultimately

✌ ✌

Any doubt that Audubon engaged in full dialogue with past ornithologists can be dispelled by paging through his own annotated copy of Alexander Wilson's *American Ornithology*, now on deposit at the Audubon Museum in Henderson, Kentucky (another, unannotated copy that belonged to Audubon is now at the Henry E. Huntington Library). Audubon's debt to Wilson was the only one acknowledged in his own lifetime, and the whiff of scandal still clings to it. Wilson's *American Ornithology* had been instantly and enormously successful from its first appearance in 1808, having answered the demand for indigenous art and science simultaneously.[11] After Wilson's untimely death in 1813, adulatory reviewers in the press clamored for a successor to continue recording native bird life.[12] Audubon was more than ready to answer the call, and his later attempts to deny this connection to his predecessor's role are largely the result of his obsession with surpassing him.[13] The achievements of Wilson and Audubon were brought into conflict initially by Wilson's Philadelphia coterie.[14] At the climax of this controversy in the 1820s, George Ord, who had

fed upper middle-class tastes for other exotic imports (including even live birds and bird skins). The bird on the page undoubtedly ranked as a possession to be displayed beside other objects of art, reflecting a colonialist appropriation of exotic native resources. In postcolonial America, this wedded a strident nationalism to bourgeois taste.

11. For Wilson's account of his intentions, see preface, *American Ornithology: or the Natural History of Birds of the United States . . .* (Philadelphia, 1808), 2:viii.

12. Clinton, review of Wilson's *American Ornithology*, vols. 5–9, for the *American Medical and Philosophical Register*, reprinted as note 17 in DeWitt Clinton, "An Introductory Discourse Delivered before the Literary and Philosophical Society of New-York, on the Fourth of May, 1814," in Keir B. Sterling, ed., *Contributions to the History of American Natural History* (New York, 1984), 84.

13. Audubon's employer at the Western Museum in Cincinnati, Dr. Daniel Drake, specifically singled out Audubon as Wilson's successor; see Daniel Drake, "Anniversary Discourse on the State and Prospects of the Western Museum Society," in *Physician to the West: Selected Writings of Daniel Drake on Science and Society*, ed. Henry D. Shapiro and Zane L. Miller (Lexington, Ky., 1970), 135. One of the most significant connections between Wilson and Audubon is Audubon's appropriation of the Wilsonian role of explorer/naturalist and of a nationalistic brand of natural history; and Audubon's contributions to the genre of natural history writing should be viewed in terms of Wilson's. A detailed comparison of their literary contributions would be useful.

14. Audubon's account of their meeting has been called apocryphal. In any event, he did not mention a network of contacts that likely led Wilson to his door. When Wilson appeared at Audubon's Kentucky shop, he was fresh from Benjamin Bakewell's in Pittsburgh. Bakewell was Audubon's wife's uncle who, four years earlier, had employed Audubon as apprentice clerk. Bakewell probably sent Wilson with an introduction (Herrick, *Audubon the Naturalist*, 1:204; and Ford, *John James Audubon*, 77). Wilson was also connected to Audubon through the Miers Fisher family, who lived near Philadelphia (and subscribed to five sets of *American Ornithology*). Fisher had been Audubon's father's American agent, shepherded Audubon *fils* in his earliest days in America, and had been involved in the complicated sale of the Audubon property in Pennsylvania; see Robert Cantwell, *Alexander Wilson, Naturalist and Pioneer* (Philadelphia, 1961), 148. Cantwell also claims that Audubon's version of his connection with Fisher was fabricated; Audubon claimed that the Fisher family intended to pair him with their daughter, and Cantwell points out that the Fishers hadn't a daughter of appropriate age.

brought the final volume of *American Ornithology* to press after Wilson's death, accused Audubon of plagiarizing Wilson's drawings.

Direct connections undeniably exist between their bald eagles, swallow-tailed kites, cerulean warblers, and more.[15] In the well-thumbed volumes of his copy of *American Ornithology*, Audubon reacted to his rejection by the Wilson partisans with his own stinging commentary. The annotations, whether simple checkmarks or more elaborate responses, appear to date from different periods, suggesting that Audubon kept the books at hand during the long period in which his own work was in production.[16] Wilson illustrated a sparrow hawk carrying off a small bird, its talons still visible. Audubon wrote, "Shew me a hawk carrying prey on the Wing and exhibiting the claws and *I* will cry out *Nature is at fault.*" He even corrected Wilson on the "words" of the American redstart's song. And, alongside Wilson's claim that he had seen a flock of whooping cranes near Louisville on the twentieth of March, Audubon bristled: "this *wonderfull* sight was granted to Mr Wilson at the request of *I* [or *J*]. Mr Wilson *then* did not believe the *existence* of *Blue Winged teal. J. J. A.*"

Audubon was even more vitriolic in his annotations to Charles-Lucien Bonaparte's continuation of *American Ornithology*, published in 1825. Bonaparte had offered to publish Audubon's "Great Crow Blackbird," which would have been the first of Audubon's drawings ever to see print. When Audubon

---

15. See Harwood, "Watershed for Ornithology," 41; Robert Henry Welker, *Birds and Men: American Birds in Science, Art, Literature, and Conservation, 1800–1900* (Cambridge, 1955), 56–57; and Clark Hunter, ed., *The Life and Letters of Alexander Wilson* (Philadelphia, 1983), 98. Ord also identified Audubon's ca. 1822 red-winged blackbird as a copy of Wilson's depiction. The birds are very similar, shown swooping downward with wings and tails spread, bills open, although they are not exactly alike: Audubon's shows articulation in the shape of the wings and the individual primary feathers, no tongue is visible, and the red and yellow patches are more rounded. He had drawn an earlier and very similar red-winged black-bird on an asclepias plant in June 1810. This drawing, too, bears a striking resemblance to Wilson's print. Although there is a foot gripping a leaf (which does not appear on either Audubon's later drawing or on Wilson's print), its wings are tighter to the body, its tongue is extended, and the wing patches are crescent-shaped—all like the Wilson. This is dated three months after the two men allegedly first met but, unlike other drawings of the period, does not cite Wilson on the inscription. In fact, this drawing precedes Wilson's plate, which was not published until 1811, in volume 4. Did Audubon misdate the drawing? Did he "borrow" an original drawing (not the plate) from Wilson? Or did Wilson "borrow" from Audubon—as Audubon claimed in other instances? Or, as seems equally plausible, were both men using past ornithological illustration, like Catesby and many other sources, as a point of departure?

16. His remarks, penciled into margins, can often be dated from handwriting and context. For example, he refers at least twice in volume 2 to warblers he shot in Philadelphia and New Jersey in the summer of 1829. In volume 8 he argues with Wilson that the black duck never ranges as far as Florida, perhaps indicating that these notes were written after his 1831 trip there. Other comments refer to the abundance of partridges and canvasback ducks at Cincinnati, Louisville, and New Orleans markets, which may suggest an earlier dating. Nearly every remark is signed with Audubon's initials or, apparently reflecting his vehemence at times, his full signature.

received his copy of the volume, he recognized that another, much inferior design had been substituted, yet still bore attribution to him. Alexander Lawson, Wilson's engraver, had simply refused to reproduce any drawing by Audubon.[17] In his marginalia in this volume, Audubon's vituperation extended well beyond his ravaged grackles. Ironically, he accused Titian Ramsay Peale, Bonaparte's chief artist, of copying from Wilson.[18] As for Lawson:

> This Wonderful hero of the Graver for ever and in defiance of all that can be said to him on the Subject—alters the Drawing. and regularly represents Both thighs and Consequently both Legs as if belonging to one Side—J. J. A. who knows a little of Neighbor Lawson.

Audubon declared his authority not only in the marginalia but also, more strikingly, in his additions to one of the printed images. On plate 8 he lightly completed in pencil the figure of an immature yellow-bellied sapsucker, where only a head was depicted.

By a curious twist, Ord perpetuated this silent dialogue by annotating *his* copy of Audubon's *Ornithological Biography* in the 1830s. Next to Audubon's entry for "Maria's Woodpecker," named for Maria Martin, the sister-in-law of Audubon's friend and later collaborator, John Backman, he commented sardonically:

> It is [a] fortunate circumstance for the credit of American Ornithology that the author of the Birds of America had not more frequent opportunities of paying such compliments or we should have had introduction to all the old maids of his acquaintance with their pretty names affixed to peckers, cocks, snipes.[19]

Wilson was not Audubon's only American predecessor. Between 1806 and 1810, Mark Catesby's images from the *Natural History of Carolina* (1731–43) seem to have influenced him particularly. Audubon's 1806 "Wood Thrush," for example, is comparable to Catesby's illustration of a related species, the

17. Alice Ford, in her 1988 biography, reproduces Audubon's original drawing (fig. 101), discovered in 1987, which proves that Bonaparte did not simply have the drawing altered but rather substituted another; Audubon was livid. For an account of the incident, see Ford, *John James Audubon*, 160–70.
18. Beside the "Rocky Mountain Anteater" (rock wren) Audubon writes, "I am sorry that Charles B. Should have taken *back ground* from Mr. Peal [*sic*] who undoubtedly must have copied it from Willson [*sic*] as his Crow Blackbird."
19. Susan Klimley, "Notes: From the Academy of Natural Sciences of Philadelphia," *Society for the Bibliography of Natural History, North American Newsletter* 2 (winter 1978), n.p.; Klimley quotes annotations from *Ornithological Biography*, 1:326.

"Fox Coloured Thrush," a brown thrasher. These two birds face in opposite directions, perhaps the result of Audubon's working with tracings. But Audubon's thrush is similar in shape, and it virtually reproduces Catesby's ineptly drawn near foot. Catesby's is posed against clusters of black cherries and Audubon's, similarly, against wild cherries. In 1808 Audubon cited Catesby's nomenclature, and later, an acquaintance recalled Audubon's stated admiration for the earlier ornithologist.[20] Clearly Audubon looked to Catesby's volumes for American source material in the years immediately preceding the publication of *American Ornithology.*

It is tempting to posit the American ornithologists as Audubon's most direct forebears. Their publications functioned as field guides: Catesby and Wilson were to Audubon what Roger Tory Peterson is to birders today. Further-more, their texts and images supported a claim, ideologically, for the impor-tance of American zoological resources. That claim was intended to rebut Buffon's controversial hypothesis that New World species were merely "degen-erate" versions of European fauna. But to separate New World from Old World zoology (as Audubon's own biased accounts might lead us to do) would be to ghettoize the Americans' achievements. Each artist/writer continually weighed the perceived new demands of American science against Old World natural his-tory. And of more fundamental importance than Audubon's specific debts to any of his predecessors is his broader visual and conceptual incorporation of these sources.

Very few of Audubon's rough working sketches survive, but those that do shed light on his ongoing dialogue with his predecessors. The 1820–21 Journal preserved at the Houghton Library, Harvard University, contains two small drawings of special interest. The first, sketched at the endpapers, is a lettered and labeled profile diagram of a songbird with its wing extended (fig. 29). Of only minor artistic note,[21] it is an almost perfect replica of the diagram in Turton's *Linnaeus,* down to the correctly labeled body parts and its T-shaped supporting branch (fig. 30). The rest of Turton's plate shows eight details of different foot types and three of types of bills, the characteristics that deter-mine Linnaeus's genera. While it has long been known that Audubon carried his copy of Turton down the Mississippi in 1820, it has not been recognized that the diagram was copied from it.

20. See William Dunlap, *History of the Rise and Progress of the Arts of Design in the United States* (1834; reprint, New York, 1969), 1:198.i

21. Davidson describes it as merely part of Audubon's self-tutelage; see Davidson, introduction, *Original Water-Color Paintings,* 1:xxix.

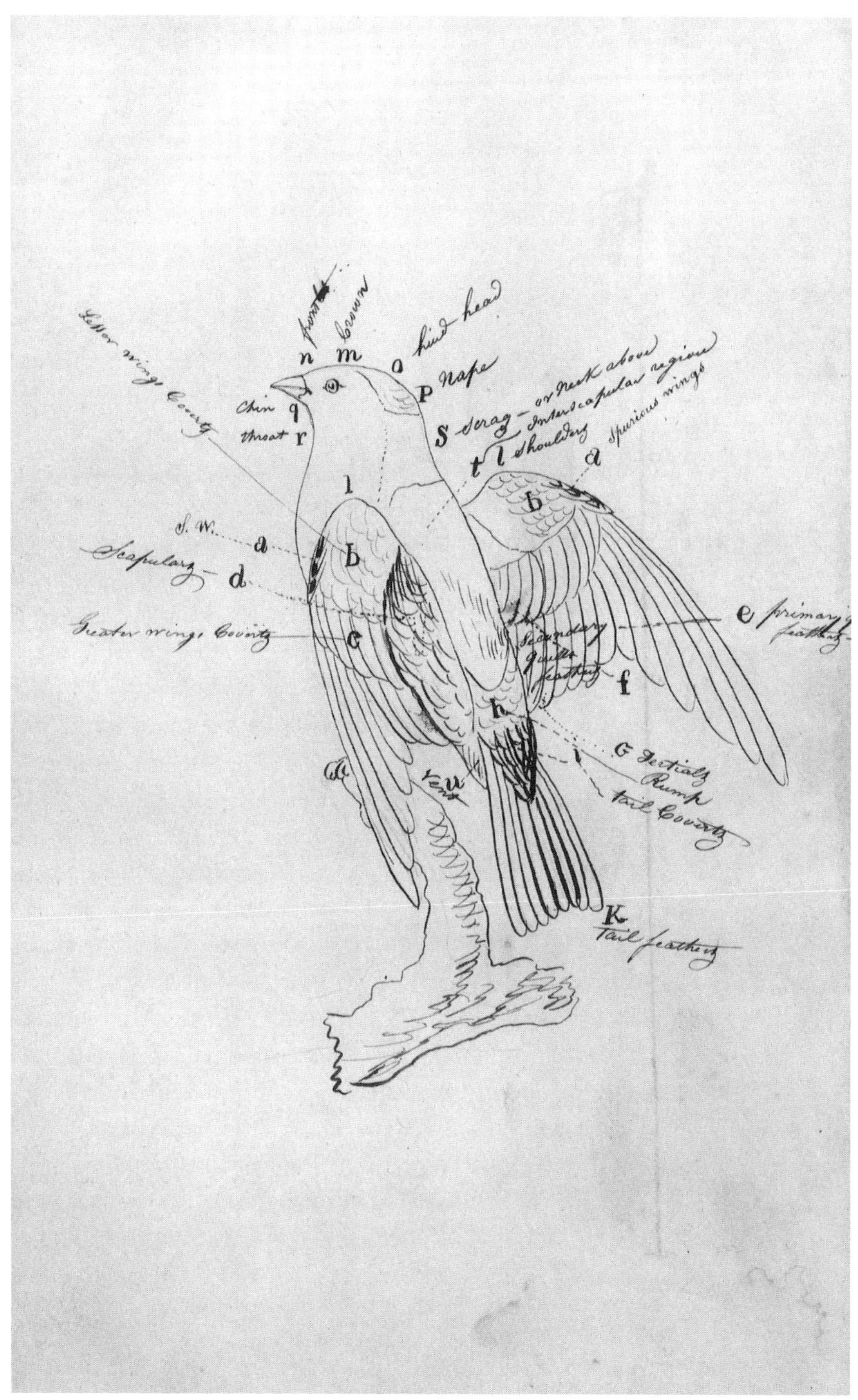

Figure 29. Diagram of bird from Audubon's 1820–21 Journal (MCZ 117F), verso of first page. By permission of the Houghton Library, Harvard University.

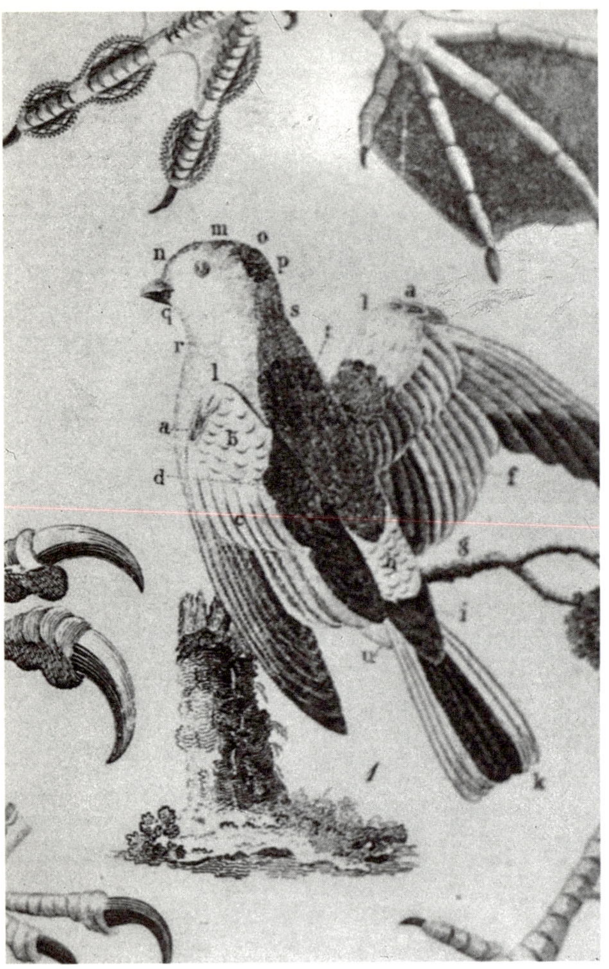

Figure 30. Diagram of parts of bird from Turton's *Linnaeus* (London, 1806), volume 1. University of Delaware Library, Newark, Delaware.

Also suggestive of the conventions of Linnaean classification are two sketches beneath an entry in the same journal. Audubon's "fin-tail" duck (ruddy duck), which he thought a discovery, is profiled on a rock, its stiff, fan-like tail and hooked bill—Linnaeus's defining characteristic—clearly visible (fig. 31). On the other hand, Audubon neglected to mention or sketch the white cheek patch, today a prime identifying field mark in all plumages. (It is faintly visible, however, in his finished drawing of the bird, pasted onto the right side of a composition of the 1830s.)[22] Beside the duck in the journal is the "imber diver," a loon, which he shot and measured. The identification and nomenclature, as well as the crucial three-toed foot, Audubon took, again, from Turton. Although Turton provided no visual models here, Audubon's bird is set in a swimming pose common in ornithological illustration: profiled with head erect, and near foot trailing, visible through transparent water. Any number of French and English illustrated waterbirds are comparable. The critical point here is that on the spot, in the Mississippi flatboat where he was examining his specimen, writing, and recording this quick sketch, Audubon was also working—and *seeing*—in the old illustrational format.

22. Ibid., 2:428.

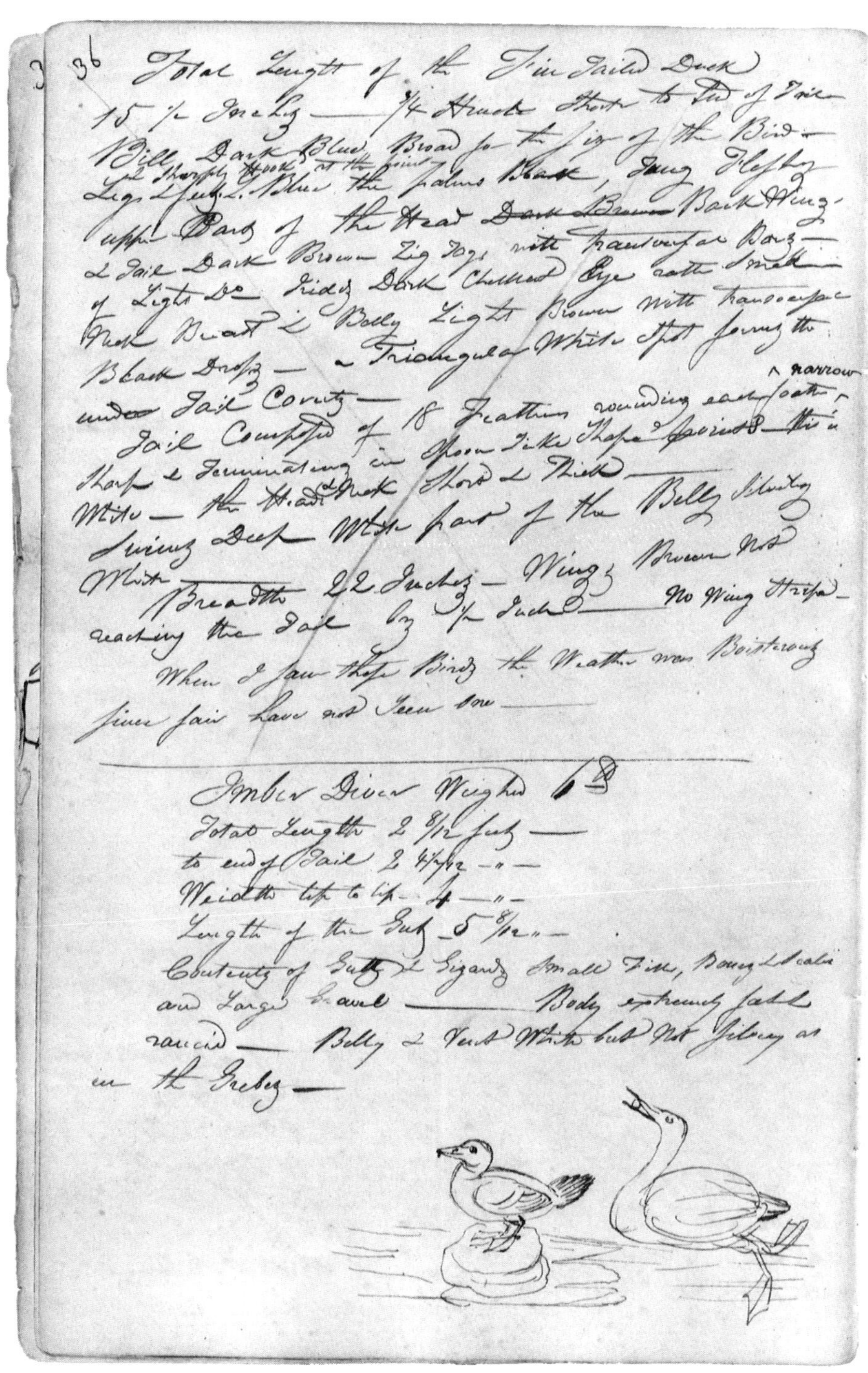

Figure 31. Sketches of loon and ruddy duck from Audubon's 1820-21 Journal (MCZ 117F), near the entry for 14 November 1820. By permission of the Houghton Library.

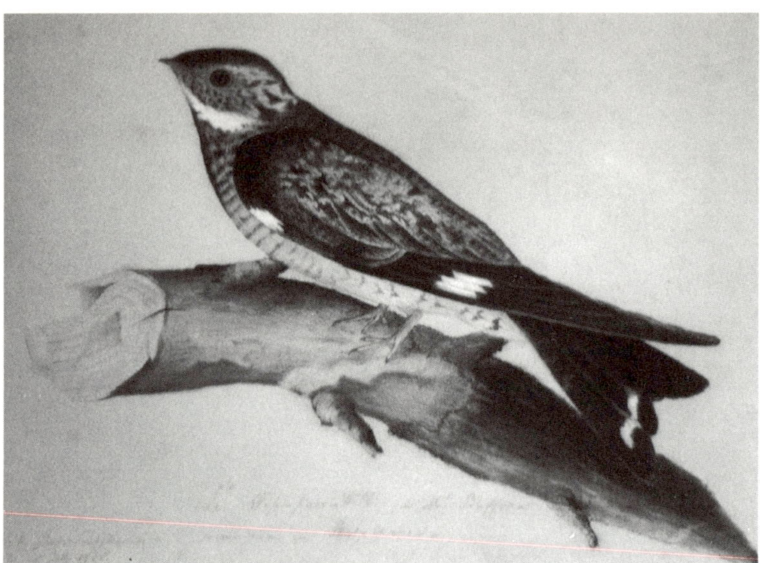

Figure 32. Audubon, drawing titled "Le Wip-poor-will de M. Buffon" (1806).
Graphic Arts Collection, Visual Materials Division, Department of Rare Books
and Special Collections, Princeton University Libraries.

One of the most handsome and adept of Audubon's early drawings, now held
at Princeton University, is inscribed "Le Wip-poor-Will. de Mr Buffon / meme
nom in Pensylvanie—Mill Grove, Pensylvania the 21 of July 1806" (fig. 32).
Depicted in rich, high-contrast brown and ochre pastels and pencil, this nocturnal insect-feeder perches along the length of a sawed-off branch. The bird is in fact
easily recognizable as a nighthawk, closely related to but not the same species as a
whippoorwill, which lacks the white wing spot. It has not been noted, however,
that the error was Buffon's, not Audubon's. This illustration—like most others in
Buffon's volumes—makes no reference to the other information that enlivens the
text: reports of its migration and native superstitions associated with the whippoorwill's peculiar call. Yet Audubon clearly followed Buffon's inaccurate textual
description of a whippoorwill's

> large black wing feathers, the first five marked by a white spot
> towards the middle of their length, and the two exterior pairs
> on the tail, similarly marked towards the tip.[23]

23. ". . . les grandes pennes des ailes noires, les cinq premières marqués d'une tache blanche vers le milieu
de leur longueur, et les deux paires extérieures de la queue marquées de mêmes vers le bout" (Leclerc
de Buffon, *Histoire naturelle des oiseaux* [Paris, 1801], 18:324).

Although Audubon was surprisingly accurate in most of his descriptions and drawings of species, a handful have not stood the test of time. At least four of his discoveries have been called "non-species."[24] New light can be shed on these and on Audubon's working process by referring again to Buffon. For example, "Cuvier's Kinglet" was drawn in Pennsylvania on 8 June 1812. It probably represented a male ruby-crowned kinglet, but Audubon shows an uncharacteristic black line encircling the front of its red crown, and its yellowish wing and tail linings are more representative of the golden-crowned kinglet. Audubon, however, would have been entirely familiar with the European kinglet species that Buffon describes in an essay immediately preceding that on the American ruby-crowned kinglet in the *Histoire naturelle*. There Buffon clearly states that the European bird's crown is bordered with black.[25] Audubon's drawing lacks the European "firecrest's" bronze neck patch but shows both its strong yellowish coloration on wing and tail base and its comparatively longer legs. The relation here is conceptual rather than visual; he reverted to a source he knew well and relied on verbal description to compensate for his own faulty memory or a damaged specimen.[26] Even lapses that we have cavalierly called Audubon's scientific errors reflect his debt to the tradition of published ornithologies.

Illustrated ornithologies also supplied Audubon with visual conventions. He inserted anatomical vignettes into his compositions throughout his career, a practice that derives from the eighteenth-century classifying ornithologists like Brisson and Linnaeus, who based their groupings of species on exterior characteristics, usually the "naked parts," feet and beaks. The illustrations in Brisson's *Ornithologie* (1760) begin with several fold-out plates of these isolated parts, carefully detailed (fig. 33). As noted, the single bird plate in Turton's translation of Linnaeus chiefly comprised vignettes of the same type.[27] A close examination of Audubon's whippoorwill drawing shows that time has nearly hidden a fine graphite frontal sketch of the bird's face. Two very round eyes

24. Harwood, "Watershed for Ornithology," 45–56.
25. *Oeuvres complètes de Buffon*, ed. M. Flourens (Paris, 1855), 7:62, 64. See also Hermann Heinzel, Richard Fitter, and John Parslow, *The Birds of Britain and Europe* (Philadelphia, 1972), 238–39.
26. Dwight follows Sir William Jardine, who revised Wilson's *American Ornithology* in 1832, in suggesting that the familiar source was a Cape May warbler (*Original Water-Color Paintings,* 2:348).
27. Even in America, zoological illustrators used the same approach well into the nineteenth century; it is found often in the *Journal of the Academy of Natural Sciences of Philadelphia*, devoted to describing new species; see for example, volume 1 of September 1817, pl. 4, by Charles Alexandre LeSueur, surely one of the lost lights of American zoological illustration. The plate of dorsal and ventral views of the "Dumeril—Shark" in this same issue is a delicate two-dimensional design that manages to suggest softness and volume, bringing beauty even to the most unlikely subject.

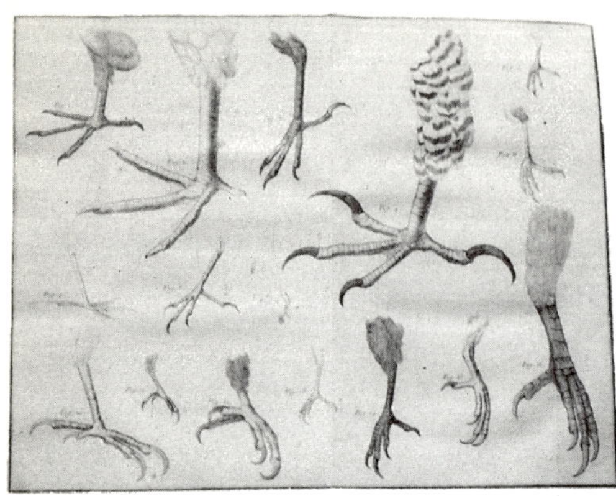

Figure 33. Details of feet, from Mathurin-Jacques Brisson, *Ornithologie* (Paris, 1760), volume 1. Division of Rare and Manuscript Collections, Cornell University Library.

with pupils are visible on the paper, browned with age, as is the inside of the mouth, opened to its fullest extent (it is too faint for reproduction here). Whippoorwills are crepuscular, and only at nightfall are their eyes fully opened. This alternative view, while static, hints at the bird's habit of catching insects on the wing. Although I have been able to establish that details of feet and beaks appear most frequently in Audubon's early drawings through 1812, the practice continued through the 1830s—even after he achieved full mastery at activating and spatially arranging his subjects.

One of Audubon's most meticulous early drawings is of a kingfisher, inscribed with the names given it by Buffon and Wilson. Along with the expected foot detail are two beautiful barred feathers, sketched nearer the bird. Audubon recorded their exact origin, sixth primary on the wing, first on the outside of the tail. He commonly noted numbers of wing and tail feathers on his drawings, reflecting descriptions like Buffon's as well as those by the classifiers. This drawing dates from 1808; however, my database shows that feathers were more commonly rendered after 1812. Likewise, feather details were commoner in European, especially French, illustrations of more modern origin. The first plate in Audebert and Vieillot's *Histoire naturelle et générale des grimpereaux et des oiseaux de paradis* (1802), printed just before Audubon emigrated from France, showed isolated bills and feathers. In an older work, the clusters of feathers that decoratively wreath the title page of George Edwards's *A Natural History of Uncommon Birds* (1743) are rendered in exacting detail, but obviously not for taxonomic purposes.

The visual convention of the old ornithologies that Audubon avowedly labored most to avoid was the static profile view. Yet the whippoorwill drawing of 1806, despite the vivid patterning and color on the bird, is so two-dimensional in contour that there is no trace of the far wing, which should be folded up over the tail. Indeed, most of Audubon's early birds are in just such profile positions, as even a fairly random selection confirms. Moreover, almost half of the 1821 birds—from Audubon's mature period—are classifiable as profiles. The pose, found as early as the turn of the seventeenth century in

Ulysses Aldrovandus's *Ornithologiae*, is as old as illustrated ornithology itself. François-Nicolas Martinet's "Manakin" and "Cock-of-the-Rock," drawn for Brisson's *Ornithologia* (1760), though shifted to a three-quarter view, retain the stasis associated with Brisson's systematics. The essentialist concept of species as fixed types parallels the eighteenth-century aesthetic of an "Ideal Beauty," a "perfect state of nature" from which all accidental blemishes are purged.[28] Martinet's symmetrical composition is, not coincidentally, constructed around a neoclassical building and it is literally underlined by a scale of measure.[29]

Of more telling interest, however, are Audubon's deliberate deviations from what we might call the norm of the static bird, viewed in profile. Half of the drawings in my database, some from as early as 1809, represent at least some activity. In 1808 Audubon drew a summer red bird leaning forward across a black locust branch (fig. 34). Though its head is profiled, the bill is open to show the characteristic mandible of a tanager, and the rest of the body is viewed three-quarters from above, tail and wings fanned. The minor movements in these and several other early drawings are only preludes to the activated "Carolina Paroquet" of 1811 (fig. 35). Although primarily profiled, the paroquet bends its head almost ninety degrees to take a nut held in its right foot. That Audubon aimed for a lifelike rendering is revealed by his note on the drawing, where he remarked in frustration the "poor imitation of colour the actual Bird being extremely Glossy and Rich."

Perhaps surprisingly, once one begins to look for models of activated birds in the traditional ornithologies, as Audubon surely must have done, examples can be found in almost every volume by every major writer. Buffon's "La Buse," a buteo hawk, shows the bird perched, but uncharacteristically set at a three-quarter view, with the head turned almost full-face to the viewer (fig. 36). (With the exception of owls, most raptors' heads were profiled to show the flesh-tearing beaks.) At the turn of the nineteenth century, Thomas Bewick printed some animated birds as well as tailpiece scenes in his books on English birds. Jacques Barraband's delicately lifelike "Le Chauve," or Capuchin bird, in François Levaillant's *Histoire naturelle d'une partie d'oiseaux nouveaux* (1801–2), turns to a three-quarter view, so that its far rounded eye projects in

28.   This notion of the ideal is articulated in Sir Joshua Reynolds's "Discourse III"; see *Sir Joshua Reynolds: Discourses on Art,* ed. Robert Wark (New Haven, Conn., 1975), 44–45. For a discussion of the essentialist concept in natural history, see Ernst Mayr, *The Growth of Biological Thought: Diversity, Evolution, and Inheritance* (Cambridge, Mass., 1982), 46, 257.

29.   The classical details found in bird pictures, from seventeenth-century Dutch and Flemish paintings to eighteenth-century French ornithological illustration, are also attributable to the high value placed on Italian training. For example, both the Dutchman Jan Baptist Weenix and the Frenchman Nicolas Robert studied in Italy in the seventeenth century; see Christine E. Jackson, *Great Bird Paintings of the World,* vol. 1, *The Old Masters* (Woodbridge, England, 1993), 78, 108.

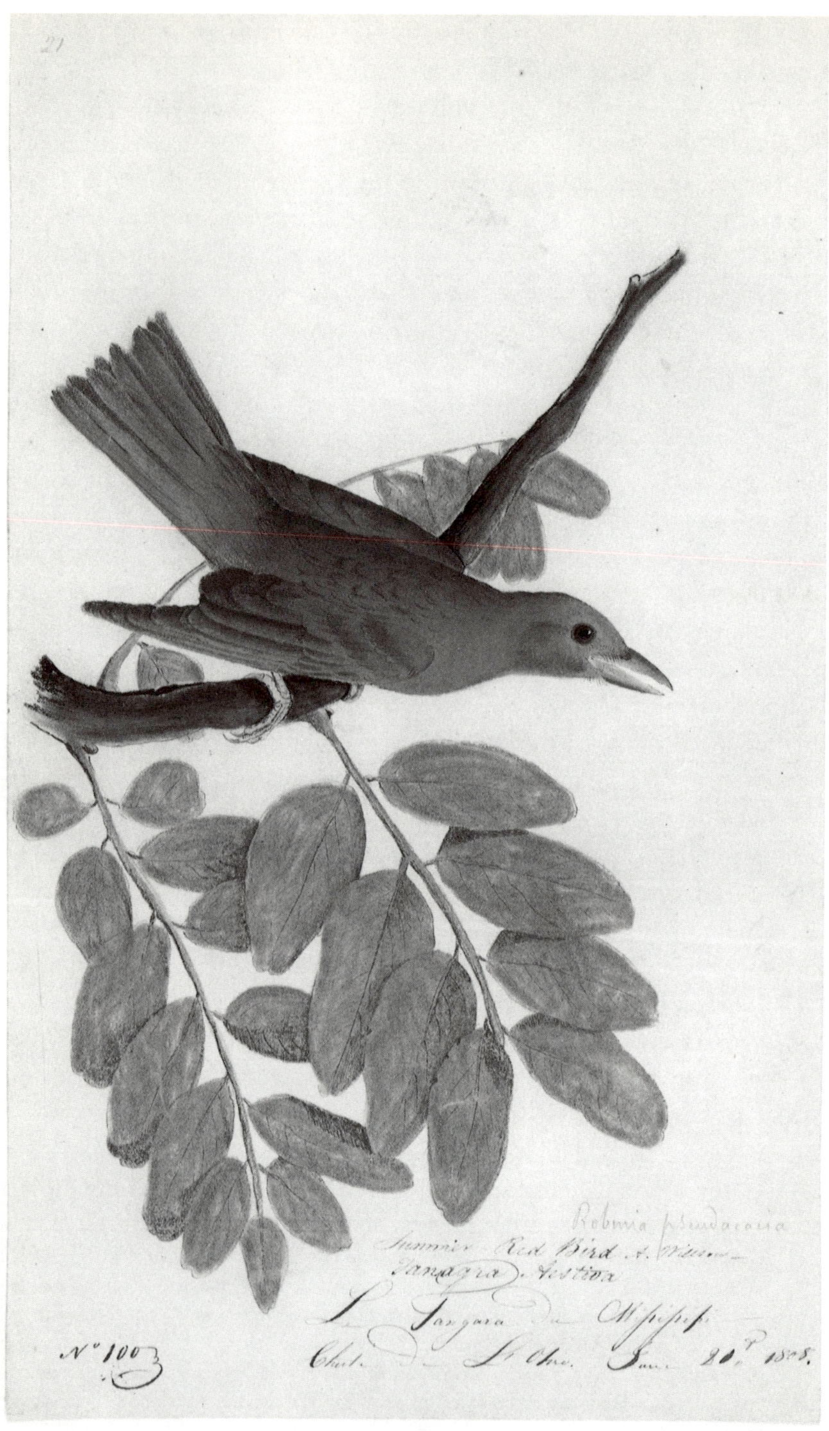

Figure 34. Audubon, drawing of summer tanager, titled "Summer Red Bird" in manuscript (1808; MS. Am 21, no. 45). By permission of the Houghton Library.

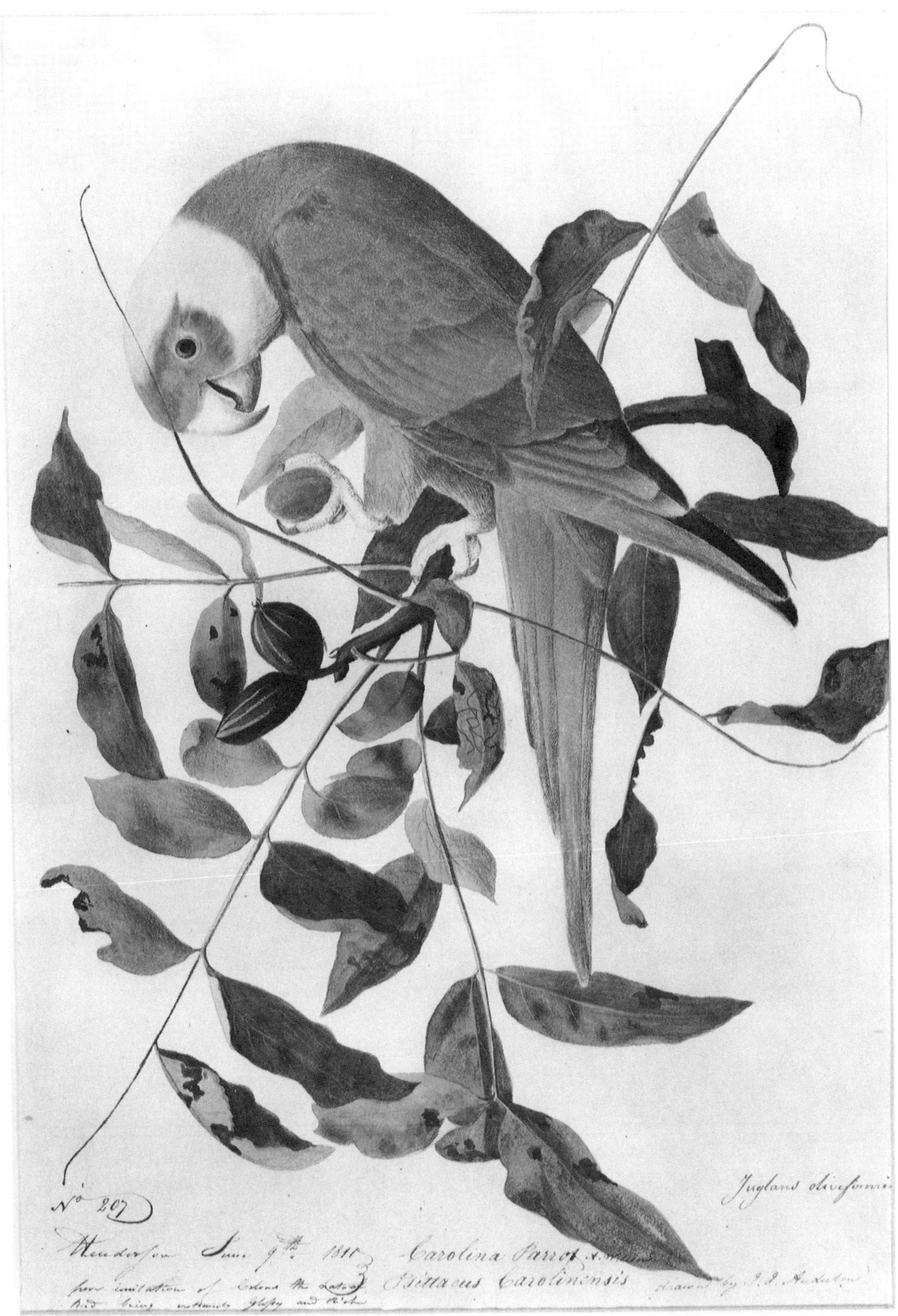

Figure 35. Audubon, drawing of Carolina parrot (1811; MS. Am 21, no. 88). By permission of the Houghton Library.

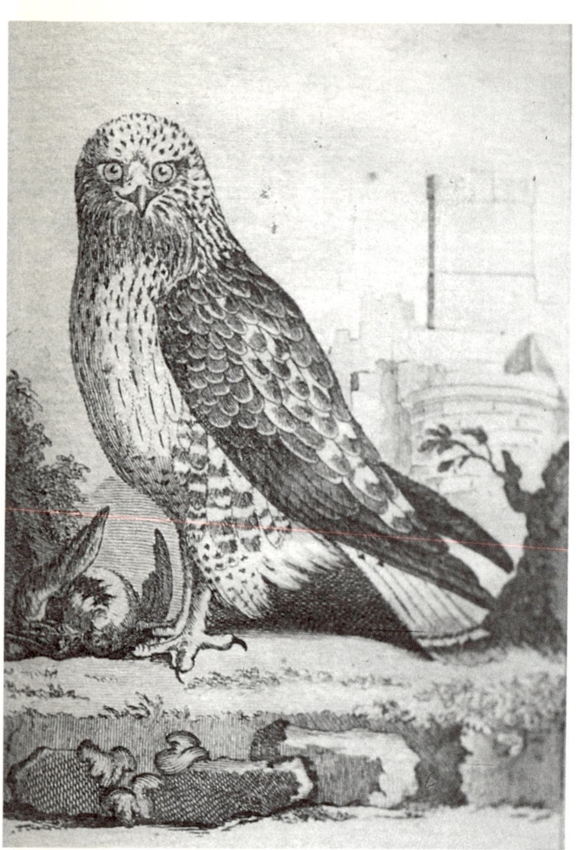

Figure 36. La buse (butteo hawk) from George-Louis Leclerc, *Histoire naturelle des oiseaux* (Paris, 1781), volume 15. Division of Rare and Manuscript Collections, Cornell University Library.

silhouette—surely an extraordinary feature by this great bird painter, at a time when drawings were typically made from dead specimens with missing or sunken eyes. Thomas Pennant's pair of Baltimore orioles pivot in a convincing, three-dimensional space (fig. 37). In a publication that targeted an American audience, the ornithological section of Rees's *Cyclopaedia* (completed in 1822), several birds—obviously borrowed from disparate sources—are jammed onto each plate (fig. 38). Significantly, however, the Philadelphia engravers tried to endow their subjects with as much energy as possible, resulting in near caricature at times. The "Condor of Magellan" clutches a wave-washed rock and looms forward, back hunched, wings poised, and beak open—animation run amok (fig. 39). The increasing attention to the habits and life-histories of birds, stressed by Buffon in particular, was itself prompted by the reports from travelers and explorers that proliferated in the eighteenth century. Allied to this, and of more direct importance to the illustrator, were developments in taxidermy that allowed the preservation and display of specimens. Private natural history cabinets were often extravagant. A treatise from 1780 suggested

> placing [specimens] on the branches of an artificial tree, which has been painted green and placed at the back of a grotto-like niche, with a small fountain, in which the water, instead of from a spring comes from a pump from a small lead cistern placed at roof level to receive rainwater.[30]

30. A. J. Desallier d'Argenville, *La Conchyliologie ou histoire naturelle des coquilles . . . Troisième édition par MM. de Favanne de Montecervelle père et fils* (Paris, 1780), 193; cited in Farber, *Emergence of Ornithology*, 49.

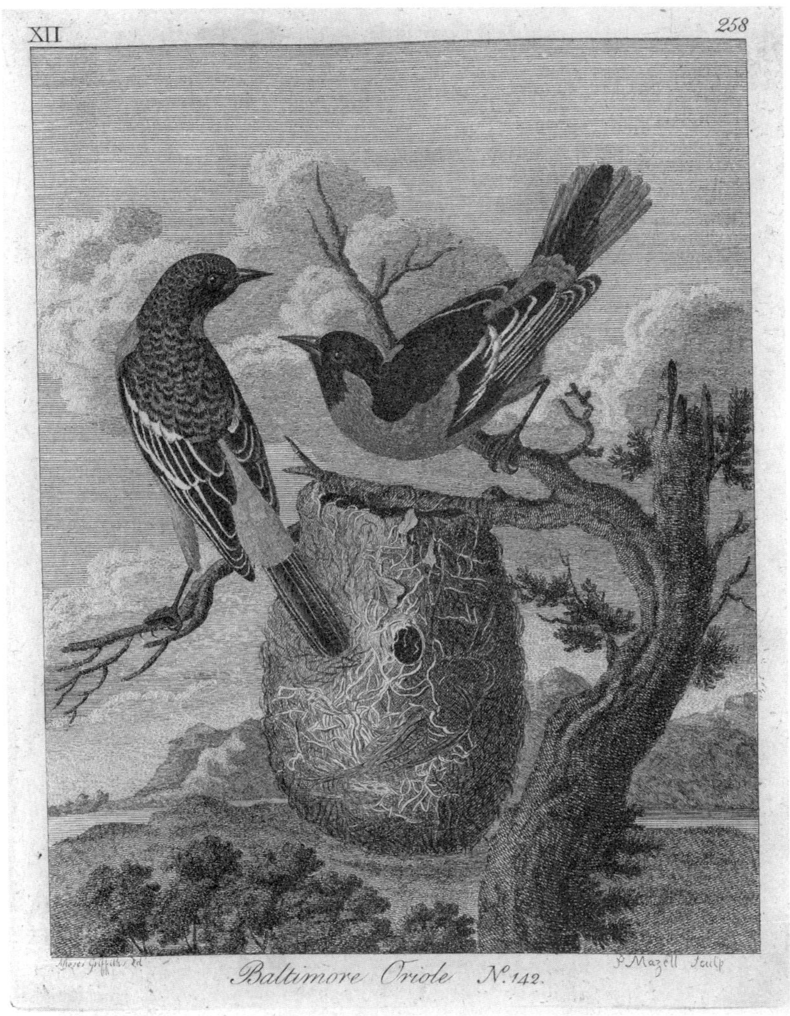

Figure 37. Baltimore orioles, from Thomas Pennant, *Arctic Zoologie* (London, 1784–85), volume 1. Huntington Library.

Illustrators, who worked increasingly with exotic birds from newly discovered lands, were hard pressed to reflect the visual splendor of the birds' own displays. Beginning late in the the eighteenth century through the end of Napoleon's reign, the French government sponsored luxury monographs on birds of exotic families and of the torrid zones.[31] Every plume was to be illustrated by the hand of a master. Barraband often raised his birds' wings or

31. Napoleon sponsored research on the arts and sciences, which rlsulted not only in propaganda but also in books of literary and artistic value. Robert B. Holtman discusses French publishing, although he does not directly consider the luxurious ornithologies produced during this era, in *Napoleonic Propaganda* (1950; reprint, New York, 1969), 31–34, 75–76.

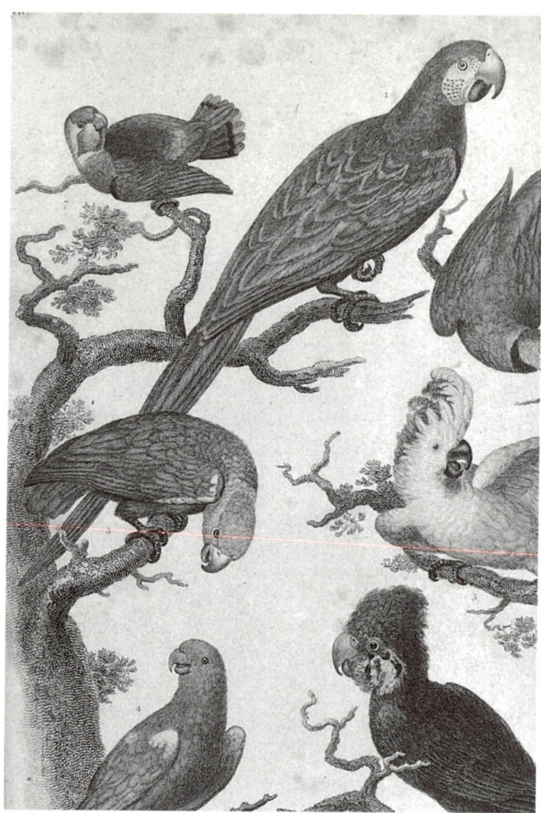

Figure 38. Parrots, from volume 5 of plates, Abraham Rees, *The Cyclopaedia; or, Universal Dictionary of Arts, Sciences, and Literature* (Philadelphia, 1810–24). University of Delaware Library, Newark, Delaware.

fanned their tails for a better view of coloration and to give some semblance of life, as he did in his designs for Levaillant's *Histoire naturelle des oiseaux de paradis* (1801–6). Audebert and Vieillot's study on birds of paradise, *Histoire naturelle et générale*, even includes a full double-page illustration of a lyrebird with its fabulous tail pattern and full-page details of individual feathers, shown near natural size. Ironically, the traditional ornithologies illustrate this display posture but fail to note its role in mating rituals—despite increased attention in the text to just such life-history.

Audubon's earliest attempts at giving his birds life, and many of his later compositions as well, should be assessed within this context. The tipped, displaying summer red bird was classed at that time with the tropical genus tanagra. Audubon could have consulted the many illustrations of American species of this genus that were found in the luxury ornithologies. His 1811 "Carolina Paroquet" can be seen in this tradition of active exotics, as can his better-

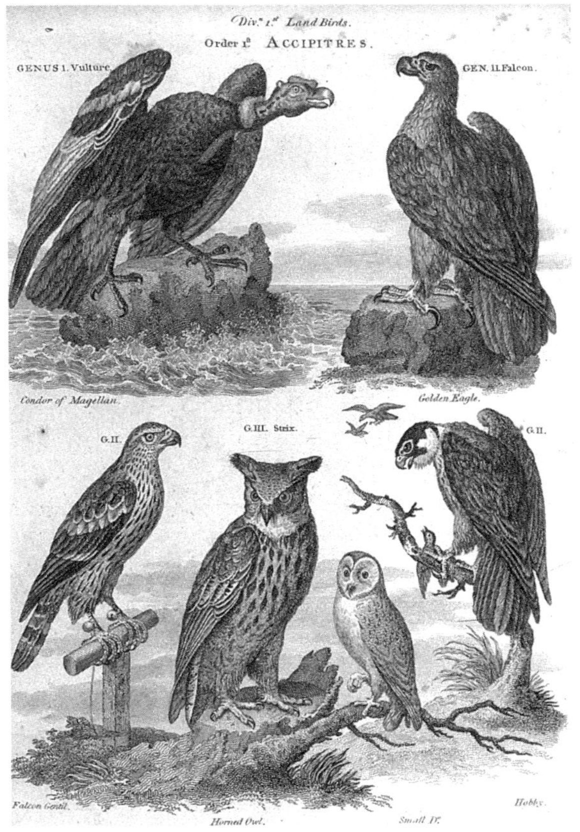

Figure 39. Condor of Magellan and golden eagle from volume 5 of plates, Rees, *Cyclopaedia*. University of Delaware Library, Newark, Delaware.

known arrangement of seven of this species, done in 1825, now twisting, climbing, and crawling in every conceivable pose.[32]

Several other types of birds had been given their own distinctive conventions of representation in the standard ornithologies. Almost without exception these, too, persisted throughout Audubon's work.

His wading birds, justifiably among his most admired, appear in feeding or preening poses—in fact a strategy that allowed them to fit, life-size, onto his paper. Distinctive as Audubon's birds are, however, this elegant adaptation of

32. Jackson mentions the tradition of painting parrots as live birds, stating that she cannot "recall an example of a dead parrot in a painting" (*Great Bird Paintings*, 65). As popular European menagerie pets, parrots must have afforded artists more opportunities to observe the living bird than did other species. Frans Snyders and other Flemish painters were allowing parrots and other colorful exotics freedom of movement by the early seventeenth century. The type was repeated even as late as the engravings for Rees's *Cyclopaedia*.

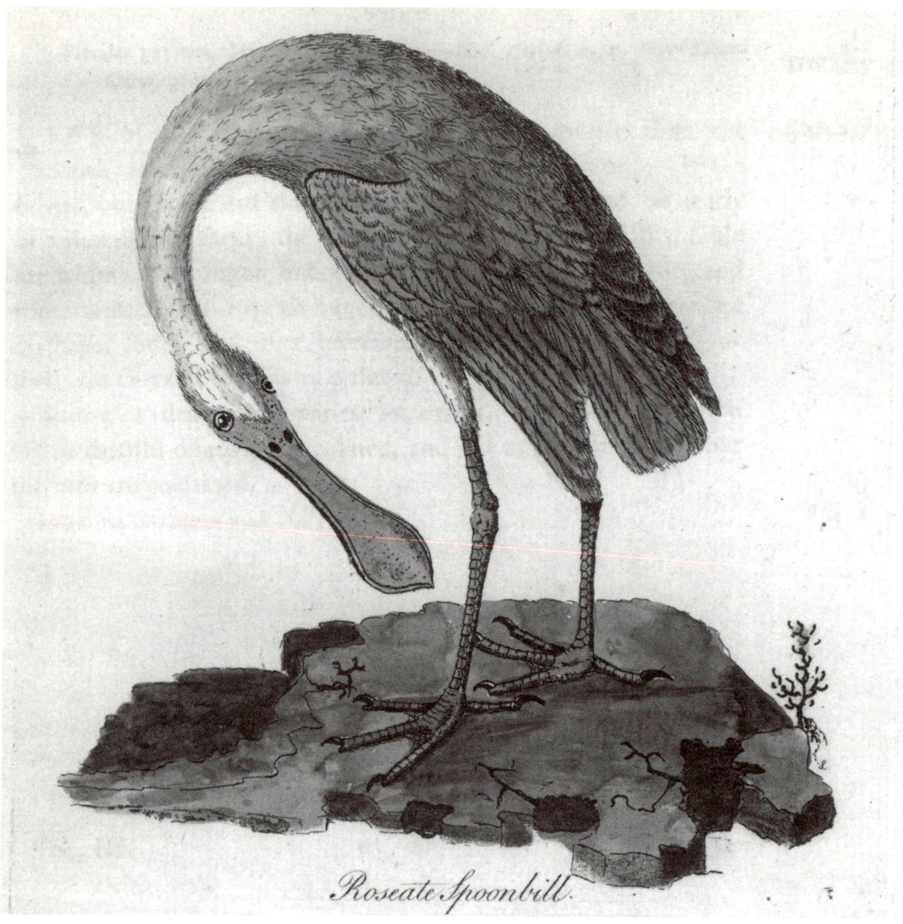

Figure 40. Roseate spoonbill, from John Latham, *A General Synopsis of Birds* (London, 1781–85), volume 4. Huntington Library.

an otherwise ungainly creature to the page was not new. Edwards's "Hooping-crane from Hudson's Bay" (*A Natural History of Uncommon Birds*, 1743–51) gracefully curves its neck to preen a secondary wing feather; and John Latham's roseate spoonbill and red flamingo (fig. 40) bends its heads downward as though feeding. Even Audubon's whooping crane (1821; fig. 41), his masterful 1831/32 roseate spoonbill, and his famous 1838 flamingo, then, revert to a proven solution from prior illustrations.[33]

33.   Taxidermists had traditionally turned specimens' heads to safeguard from accidental snagging and breakage (Jackson, *Great Bird Paintings*, 124).

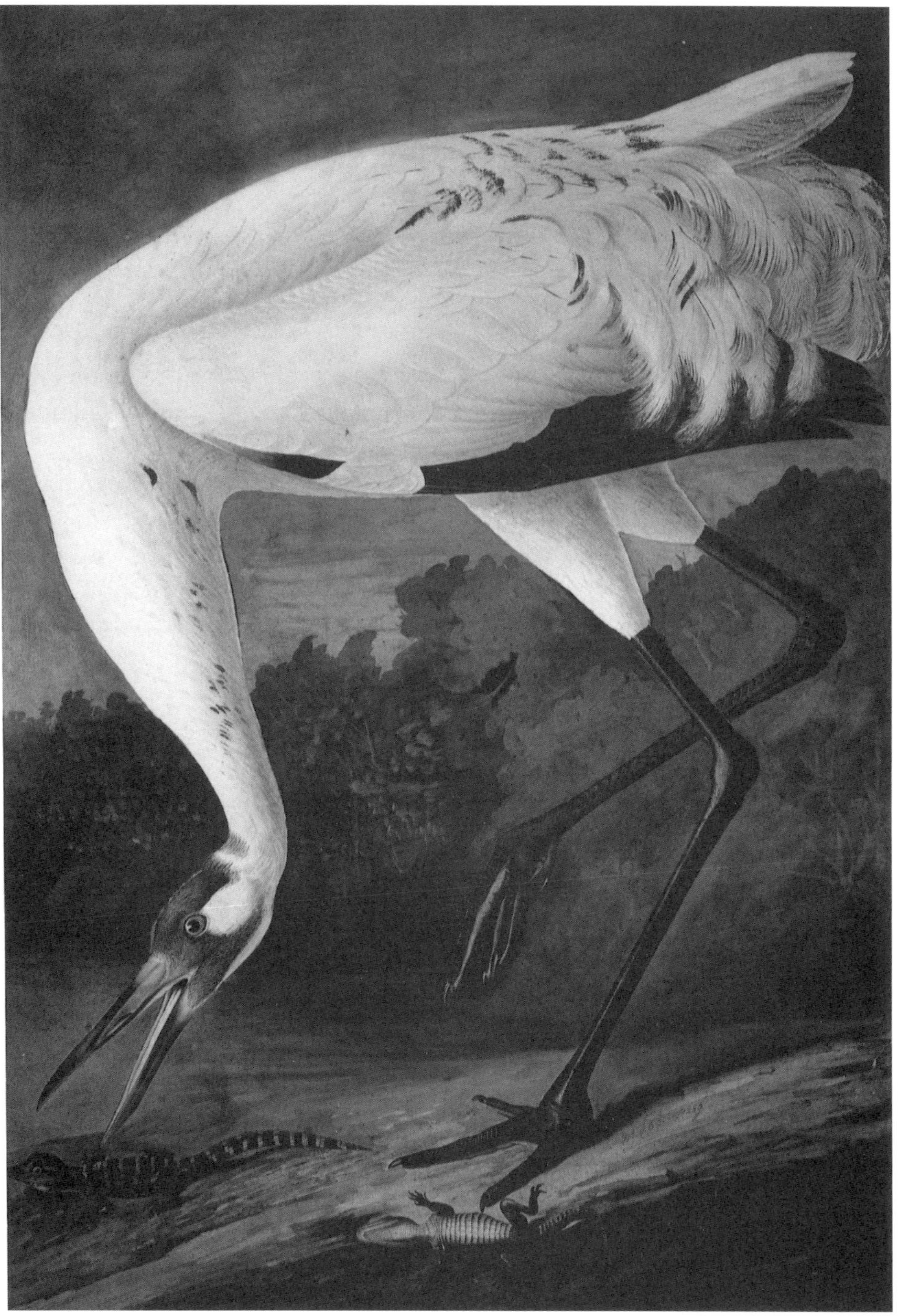

Figure 41. Audubon, Whooping crane (1821); watercolor, 94.7 x 65.2 cm. Collection of the New-York Historical Society.

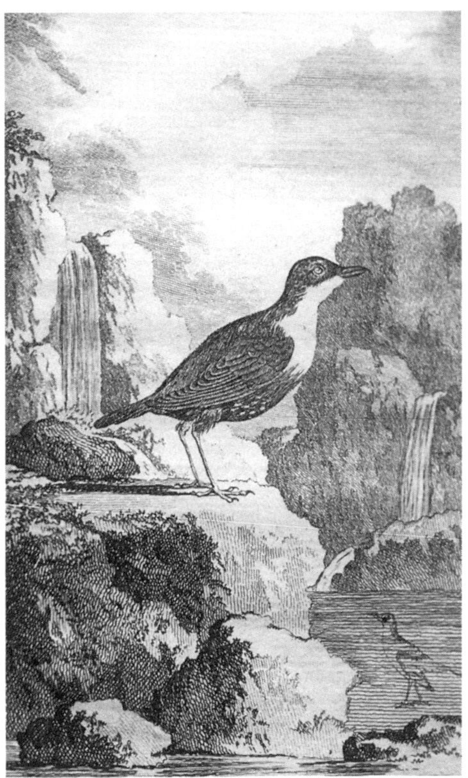

Figure 42. Merle d'eau (dipper), from George-Louis Leclerc, *Histoire naturelle des oiseaux* (Paris, 1781), volume 15. Division of Rare and Manuscript Collections, Cornell University Library.

So that every feature of a waterbird could be seen clearly, it was common to use the device of transparent water. No illustration is more peculiar than the water-ousel, or dipper, illustrated in an early edition of Buffon, where one bird is wholly visible atop a rock and below it a second can be seen entirely immersed, strolling along the bottom of the river (fig. 42). Swans were traditionally shown swimming with wings arched like wind-filled sails and one foot paddling in clear water. Bewick's mute swan is characteristic (fig. 43); and the mute swan in Rees's *The New Cyclopaedia* appears so similar that perhaps it derived from Bewick. Audubon's trumpeter swan of 1821 or 1822 is similar, though its visible near foot is seen against the marsh grasses. His 1837 trumpeter returns to the water, foot still extended (fig. 44). To deal again with the long neck, Audubon has adopted the expediency of curving it backward as though the bird is about to preen a wing. This device actually places it closer to Edwards's "Wild Swan,"[34] which, like Audubon's, appears awkward at the curve of the neck and at the juncture of the neck with the body, a danger inherent in departing from the usual view (fig. 45). (A problem shared by Audubon and Edwards was that, unlike the tame European mute swan of ponds and parks, wild swans like the trumpeter do not usually hold their necks in the expected S-curve.)

34. Edwards's illustrations especially seem to be descendants of the seventeenth-century Dutch and Flemish genre of bird paintings in the tradition of Melchior de Hondecoeter, with which they share bulging contours and contorted, open-mouthed poses. The goose that is a central element in Jan van Oolen's composition, *A Brace of Mallard, a White Goose, a Curlew, a Snipe and Other Birds by a Plinth with a Coastal Landscape* (see Jackson, *Great Bird Paintings*, 128), is directly comparable to the Edwards bird; and any number of other Dutch examples would have been available to British artists through popular engravings. Edwards, however, obeyed the dictates of eighteenth-century science by isolating his specimen—except for a perfunctory submerged "waterscape" and the addition of an alternative head and bill detail, and by allowing an unobstructed view of the critical webbed foot. I am indebted to Bruce Robertson for sharing his insights connecting Flemish and Dutch bird paintings with ornithological illustration.

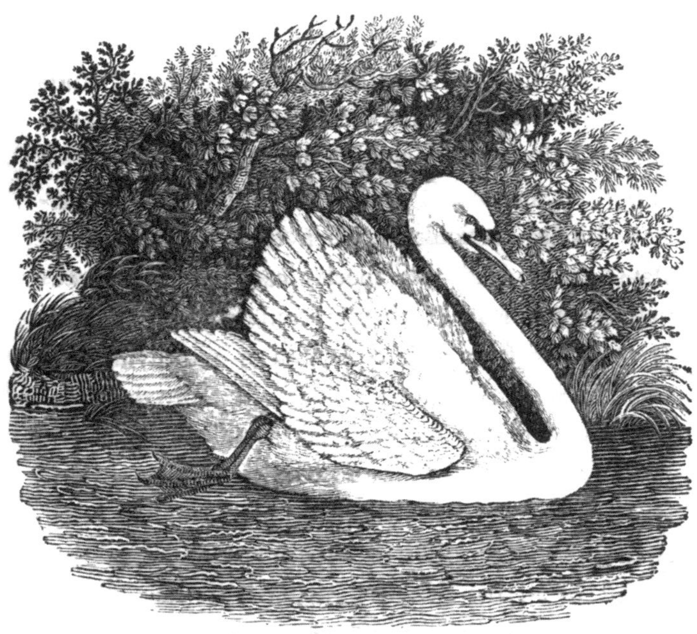

Figure 43. Mute swan, from Thomas Bewick, *History of British Birds* (Newcastle, England, 1797–1804), volume 2. Huntington Library.

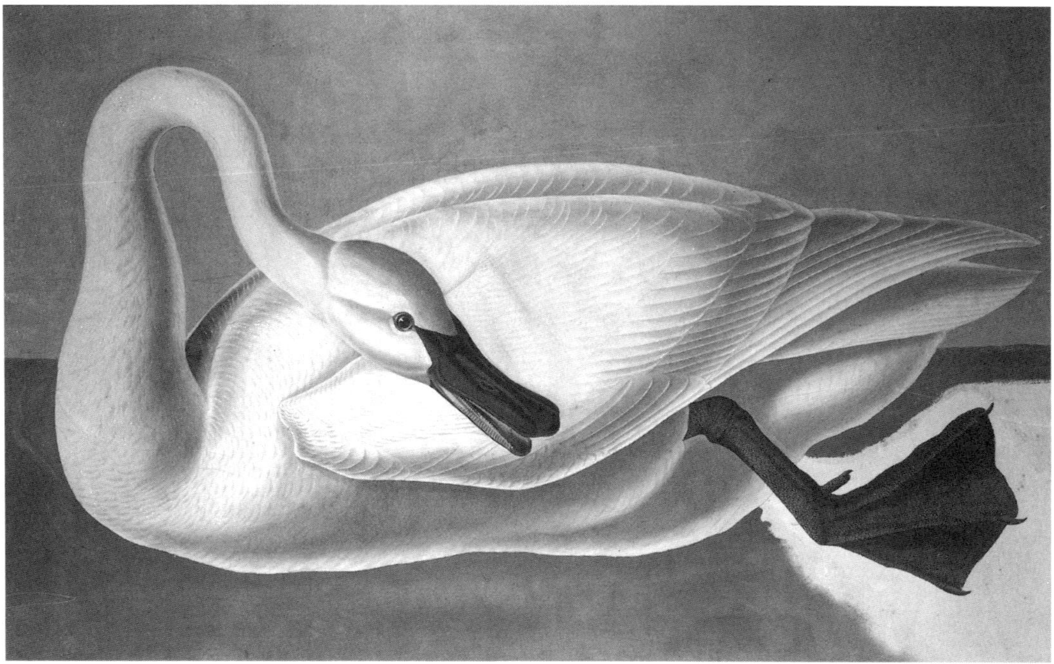

Figure 44. Audubon, Trumpeter swan (1837); watercolor, 58.5 x 95.9 cm. Collection of the New-York Historical Society.

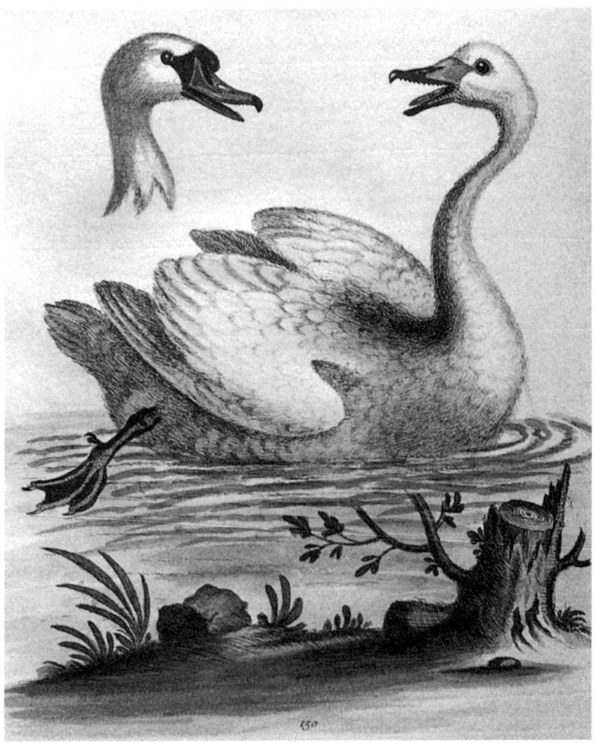

Figure 45. Wild swan, from George Edwards, *A Natural History of Uncommon Birds* (Amsterdam, 1772–81), volume 3. Huntington Library.

Northern seabirds presented a challenge to the illustrator. Many species only take to land for their brief breeding season, choosing sheer cliffs or small rocky islands as the best defense against marauding gulls. Vigorously defended territories of an individual pair are necessarily tiny, perhaps two to three square feet, and sometimes several thousand breeding birds share the same rocks. As should be obvious from even a cursory overview, eighteenth- and nineteenth-century illustrations did not uniformly ignore habitat, and some birds, like Buffon's dipper, are remarkably well integrated with their surroundings. Integration of a single type specimen—that is, a bird deemed absolutely typical, the epitome for all time, of an unchangeable species—within a large, unruly flock, however, was another matter. Compositions crowded with birds might threaten to undermine the desired perception of natural order and the primacy of the type specimen over random individuation; and such compositions would be costly and time-consuming. Instead of massing them shoulder-to-shoulder, then, artists showed one or two individuals backed by the sea and a distant "bird

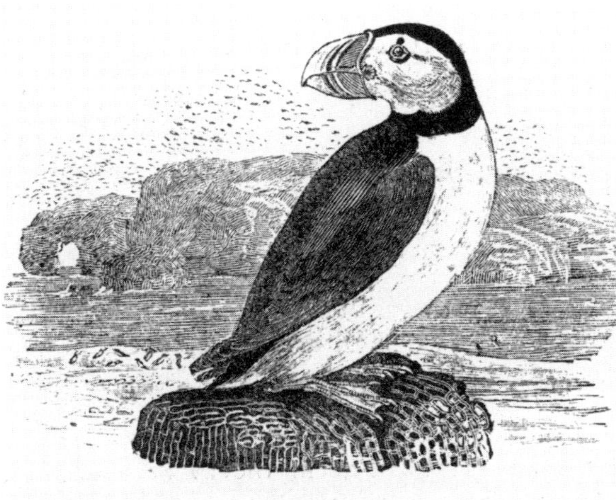

PUFFIN.

MULLET, COULTERNEB, SEA-PARROT, POPE, OR
WILLOCK.

Figure 46. Puffin, from Bewick, *History of British Birds*, volume 2. Huntington Library.

rock" with dots or parallel hatch lines to represent the swarms of soaring adult birds. This was Bewick's approach for his puffin (fig. 46). Audubon's 1833 gannets (fig. 47) are also flanked by the traditional "bird rock"; they are shown on their exposed, barren nest site and, true to tradition, isolated from the rest of the flock (fig. 48). This convention does, however, highlight the true innovation Audubon achieved elsewhere, in his multiple bird compositions. Furthermore, the 1833 illustration suggests his conceptual shift away from type and toward the individual.

Audubon's stylistic development and handling of materials can also be discussed profitably in terms of the old ornithologies. Two aspects of the physical presentation of his drawings are of special interest: his experimentation with gold paint and his decision to reproduce his specimens life-size.

In a few instances Audubon used gold paint for feather details or iridescence.[35] The most notable example is the turkey cock of about 1825; but even an 1811 killdeer is touched lightly with gold color on the tail. French illustrators

35. See Reba Fishman Snyder, "Complexity in Creation: A Detailed Look at the Watercolors for *The Birds of America*," in Blaugrund and Stebbins, *Watercolors for* The Birds of America, 62. Snyder notes that the metallic paint was used in paintings of the wild turkey, mallard, and hummingbirds now in the New-York Historical Society collection. Because the paint has oxidized it cannot be gold, but probably brass or bronze based.

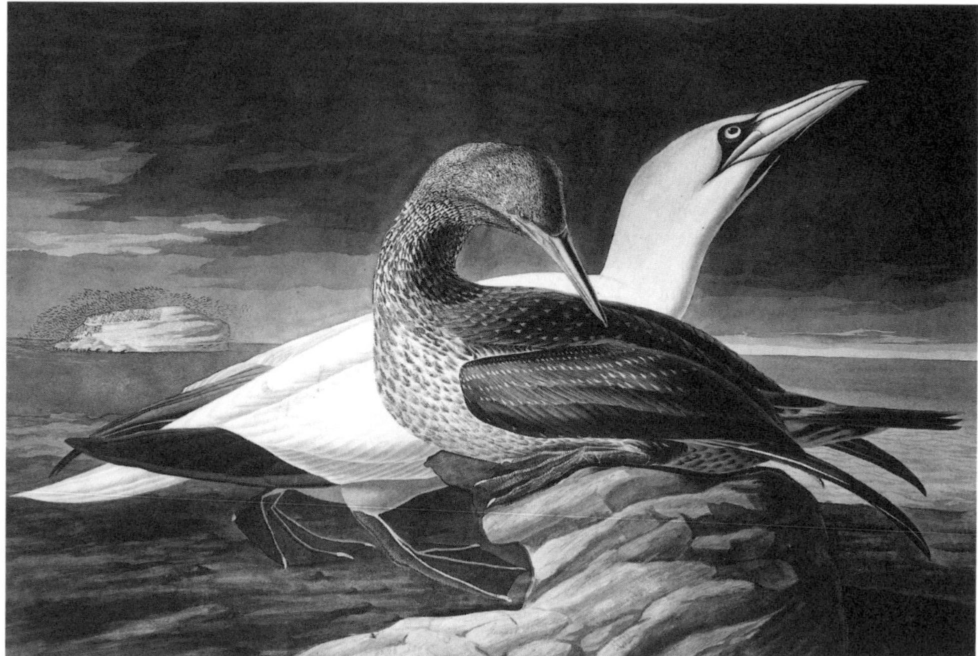

Figure 47. Audubon, Gannets (1833); watercolor, 62.8 x 95.6 cm. Collection of the New-York Historical Society.

of the early nineteenth century, beginning with Audebert and Vieillot's *Oiseaux dorés* of 1802, had turned to gilding to heighten the metallic reflections of tropical birds. Gold was used as underpainting, highlighting (and at times even for plate numbers). At the same time, the process raises issues implicit in the French productions. In the colonial period, not only was the fine printed book a mark of wealth,[36] but the exotic bird itself, whole or in part—as pet, specimen, or elegant illustration—was also a desirable object, uniquely qualified by its sensual softness, contour, and color to serve as a sign of opulence.

Audubon was also far from being the first artist to represent birds life-size. Even in works where the images were reduced, artists often attempted to show scale. Martinet placed a measured bar at the base of his plates. The puffin and razorbill in Edwards's *Gleanings of Natural History* share the page with line drawings of heads of the same species at full size. Wilson used the same tactic with his black vulture and turkey vulture. Life-size renderings were almost a universal aim of the luxurious French ornithologies. The prospectus for Levaillant's *Histoire naturelle des oiseaux de paradis* offers the viewer the most

36. For a discussion of printing conventions and natural history illustrations, see Blum, introduction, *Picturing Nature.*

seductively colored birds, "and in order that the amateur should want for nothing, each species is rendered natural size."[37] (Audubon later purchased this volume for his own library.) Vieillot boasted in the "Avertissement" at the beginning of his 1807 natural history of American birds, *Histoire naturelle des oiseaux de l'Amérique septentrionale*, that his life-size illustrations surpassed all previous works on North America.[38] Audubon, then, once again had taken his cue from the earlier ornithologies, and moreover, with Wilson he shared a chauvinistic commitment to refuting Buffon's claim that American fauna were but dwarfed and degenerate races of European species. And like Levaillant and Vieillot, he strove to produce a luxury commodity. Although he was advised by friends and by publishing professionals in England to reduce costs by reducing scale, he was correct in foreseeing that producing *The Birds of America* as a double-elephant folio would in fact enhance the marketability of his eloquent drawings among the monied classes that he solicited as patrons.

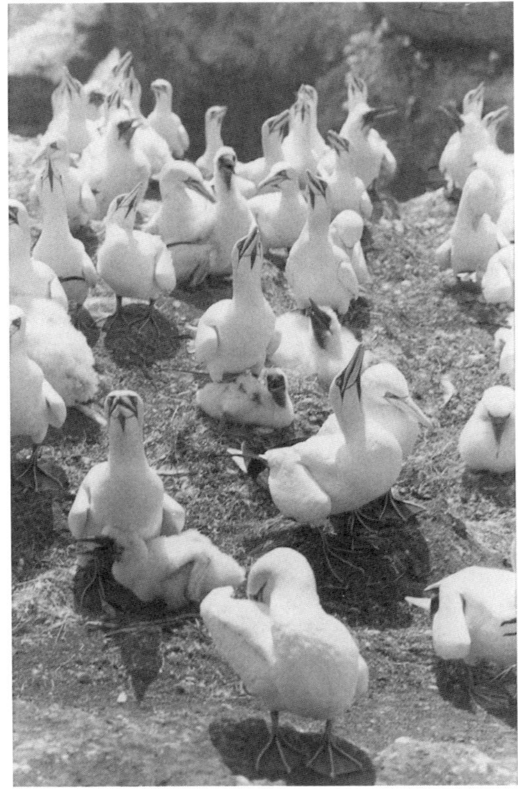

Figure 48. Gannet colony, Ile Bonaventure, Quebec. Author's photograph.

It would be a mistake to assume, of course, that for every convention Audubon borrowed he retained its subtext. Yet he was resourceful, was better informed than has been generally acknowledged, and was a master at manipulating his public.

37. " . . . et pour ne rien laisser à désirer aux amateurs, chaque espèce est figurée de grandeur naturelle" (prospectus for François Levaillant, *Histoire naturelle des oiseaux de paradis* [1806], quoted in René Ronsil, "L'art français dans le livre d'oiseaux," *Mémoires du Muséum National d'Histoire Naturelle*, ser. A [Zoologie] 15 [1957]: 38).

38. Jean Louis Pierre Vieillot, avertissement, *Histoire naturelle des oiseaux de l'Amérique septentrionale* (Paris, 1807), 3. Philip R. St. Clair, in his preface to Vieillot, *Songbirds of the Torrid Zone*, trans. Eve Jentoft (Kent, Ohio, 1979), notes that Vieillot "was among the first to illustrate his works with life-size portraits, an innovation that doubtlessly influenced John James Audubon as he conceptualized his monumental *Birds of America*" (p. xvii).

Finally, there is one raptor image by Audubon with unexpected origins. Of his "Bird of Washington" he wrote,

> If ornithologists are proud of describing new species, I may be allowed to express some degree of pleasure in giving to the world the knowledge of so majestic a bird.[39]

Audubon was convinced this was a distinct species; in fact it is an immature bald eagle.[40] The watercolor drawing for plate 11 of *The Birds of America*, inscribed New Orleans 1822 (fig. 49), may be a copy of an earlier pastel,[41] since Audubon recorded taking a specimen in 1816 and seeing his last live example in 1821. The eagle is perched on a rock (which, Audubon noted in his *Ornithological Biography*, was this bird's nesting habitat), back three-quarters to the viewer, its stance vertical. The curve of its shoulders is sweeping and exaggerated. The design and handling of some drawings dating from 1814 to 1816 are more relaxed and assured than in the stilted "Bird of Washington"; its archaic quality is odd even in light of his reliance on illustrational conventions.

Audubon's "Bird of Washington" belongs to a venerable tradition following the classic Roman eagle. The type was utilized in medieval falconry treatises and was adapted to the northern baroque genre of animal painting. It recurs in raptor illustrations from Ulysses Aldrovandus's encyclopedic *Opera Omnia* (1599–1603) through ornithologies of the early nineteenth century. For instance, the exquisite golden eagle that Barraband drew for Marie Jules-César de Savigny's ornithology *Système des oiseaux de l'Egypte* (1810), published pursuant to Napoleon's Egyptian campaign, carried a substantial iconographic weight on its noble shoulders. Few Frenchmen needed to have the connection between this "scientific" illustration and Napoleon's imperial eagle pointed out to them.[42] Although Audubon's bird resembles Barraband's in crisp contour, their shared debt is to the neoclassical tradition in general.

Still, the "Bird of Washington" is, with the exception of borrowings from Wilson and Catesby discussed above, the only one of Audubon's drawings with an identifiable source. Its specific origin was a plate dated 1806 in Rees's *New*

---

39. Audubon, "Bird of Washington," *Ornithological Biography*, 1:65.
40. Ibid., 58–62.
41. *Orginal Water-Color Paintings*, vol. 2, pl. 228.
42. Audubon titled his own stiff and heraldic merlin for *The Birds of America*, pl. 75, "le Petit Caporal," honoring Napoleon.

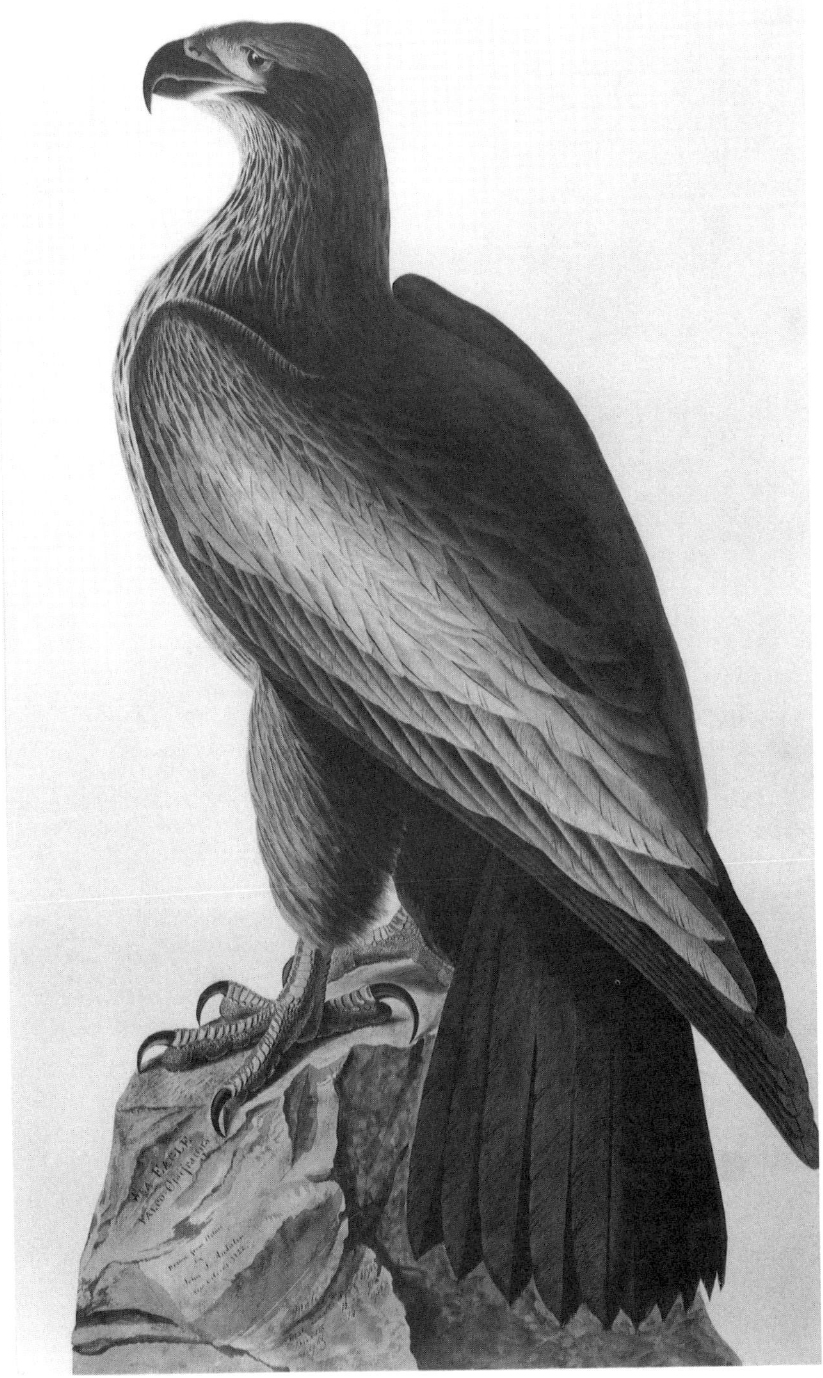

Figure 49. Audubon, "Bird of Washington" (1822); watercolor, 96.8 x 65.1 cm. Collection of the New-York Historical Society.

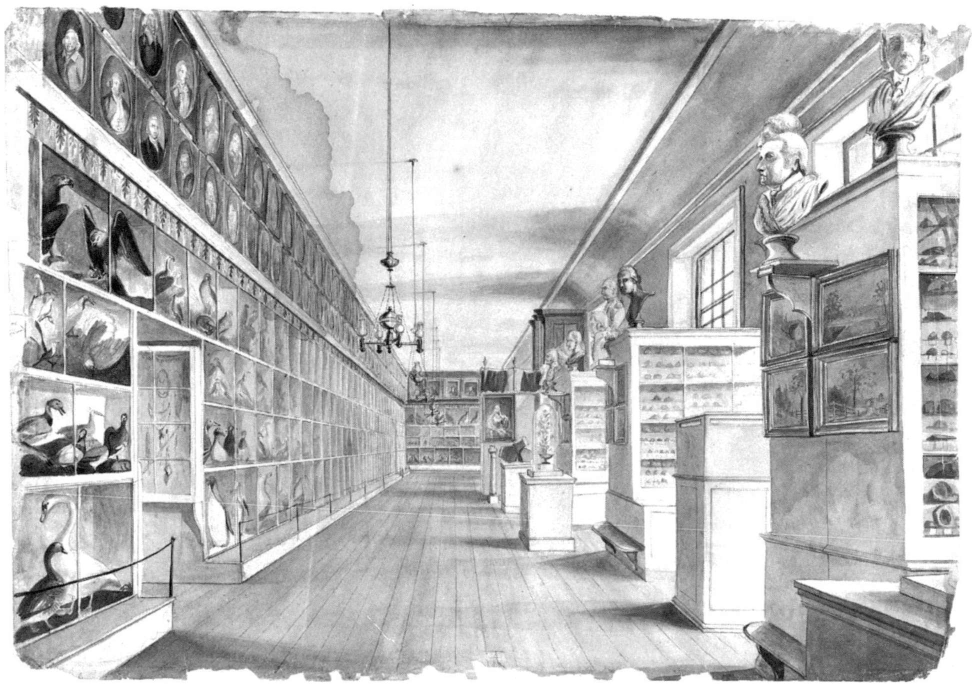

Figure 50. Titian Ramsay Peale and Charles Willson Peale, *The Long Room* (1822); 35. 6 cm x 52.7 cm. Detroit Institute of the Arts, Founders Society Purchase, Director's Discretionary Fund.

Cyclopaedia, showing a golden eagle on its rock opposite the condor of Magellan (see fig. 39). The birds apparently were reproduced after specimens on display in cases lining the left wall of the Long Room in the Peale Museum, as identifiable from an 1822 watercolor (fig. 50). That Audubon relied on either Peale's displays in Philadelphia, or the *Cyclopaedia*, published in the same city, is a notion new to Audubon scholarship. Evidence points to the published plate as his direct source. The golden eagle is perched on its outcropping precisely as is Audubon's bird, with the same curve to the wings and wingtip overlapping tailfeathers, the same angle of stance, contour of the head and beak, furrowed brow line, and even an identical highlight in the eye. The feet on Audubon's bird are more complete, a consequence of his emphasis on the importance of the claws in species identification. He clearly shows ten tailfeathers in the drawing, where he specifies twelve in his text;[43] but there are ten, though asymmetrically arranged, in the Rees print. Audubon's bird faces to the left, as does the eagle in the *Cyclopaedia*, whereas the original mount

43. Audubon, "Bird of Washington," *Ornithological Biography*, 1:63.

faced to the right to bracket the grouping in its case. These discrepancies suggest, then, that Audubon borrowed from the illustration. The *Cyclopaedia* would have been easily available to Audubon on any of his several business or family trips to the Philadelphia region; or perhaps it was obtainable in the library of a friend in Kentucky or New Orleans. It was definitely on the shelves of the Cincinnati public library when Audubon worked there preparing his own museum displays in 1820, when he would have been especially predisposed to search out models.

On one hand, it is ironic that Audubon relied on a published, immobilized, and formulaic eagle as the matrix for the bird he was proudest to discover. He could have chosen a more mobile model, for raptors pictured with prey had always been an acceptable alternative convention. On the other hand, the hierarchic Roman profile would have been well suited to underscore his claim of cataloguing a new species, and to emblematize his find as a national image.

This near-plagiarism was not seized upon by Audubon's Philadelphia enemies in his own time, although they had the originals near to hand. Perhaps it held less interest for them because it bore no relation to Wilson's illustrations and did not impinge on his primacy in any way. Precisely because Audubon relied on a format so grounded in both traditional imagery and ornithological illustration, his specific source remained masked.

John James Audubon's reputation far surpassed that of any other ornithological illustrator, and his bird drawings were as accomplished as he and his promoters claimed. But, contrary to his self-styled legend, his education was not achieved in the wilderness, nor did he simply use his early reference tools to catapult his own work far beyond their range. Rather, if we can borrow Buffon's metaphor, Audubon does not stand at the apex of some attenuated Chain of Being. He is one glittering strand in a broad and dense woof.

Marywood University

The Poetics of Geologic Reverie:
Figures of Source and Origin in Samuel Seymour's
Landscapes of the Rocky Mountains

———————————————— KENNETH HALTMAN

> Une fois de plus on peut se convaincre que voir de près c'est
> s'interdire de rêver loin.
>
> —Gaston Bachelard[1]

In 1782, Benjamin Franklin wrote in a letter to a French colleague:

> I know not whether I have expressed myself so clearly, as not to get
> out of your sight in these reveries. If they occasion any new
> enquiries, and produce a better hypothesis, they will not be quite
> useless. You see I have given a loose to imagination; but I approve
> much more your method of philosophizing, which proceeds upon
> actual observation, makes a collection of facts, and concludes no
> farther than those facts will warrant. In my present circumstances,
> that mode of studying the nature of this globe is out of my power,
> and therefore I have permitted myself to wander a little in the wilds
> of fancy.[2]

Portions of this essay were presented in papers at Bryn Mawr College and the Yale University Art Gallery in 1992, the American Studies Association meeting in 1993, and the Missouri Conference on History in 1994. I have benefited from the comments of Stephen Z. Levine, Angela Miller, Bruce Robertson, Roger Stein, and Bryan Wolf; my thanks to Erin Wright for a careful critical reading.

1. "Yet again it seems to be the case, that to look closely may be a means not to dream from afar" (Gaston Bachelard, *Fragments d'une poétique du feu*, written in 1961 and published posthumously [Paris, 1988], 67).
2. Benjamin Franklin to the abbé Jean-Louis-Giraud Soulavie, 22 September 1782; in *Transactions of the American Philosophical Society* 3 (1793): 4–5. This letter is reprinted in its entirety, with some orthographic changes, in *The Complete Works of Benjamin Franklin*, ed. John Bigelow, 10 vols. (New York and London, 1887), 8:187–92.

This apology for indulgence in geologic reverie followed an uncharacteristically fanciful passage, in which Franklin—that exemplar of scientific empiricism— attributed irregularities in the surface of the earth to the action of waves produced in its "internal ponderous fluid" by underground explosions of water and fire. The sudden violence of such an explosion, he mused, would suffice not just to uplift the incumbent earth above but also to impress the fluid below, creating a wave that might "run a thousand leagues lifting and thereby shaking successively all the countries under which it passe[d]."

The scientific merits of this brief meander "in the wilds of fancy" interest me less than does the twofold case that Franklin made for it. Given the choice, he explained, he would more approve a "method of philosophizing" based on "actual observation"; but in his "present circumstances"—he was in Paris to conclude a treaty with Great Britain—he had little opportunity for geologic let alone geogenic study in the field. However, in the paragraph quoted he adduces a further, positive rationale for giving "a loose to imagination": that there might be some explanatory value in reveries, if they "occasion any new enquiries and produce a better hypothesis" in the end.[3]

As might be expected from Franklin, both of these justifications for speculation are ultimately practical ones—a recuperation of imagination on the grounds of common sense. In fact, I would suggest, the tension that his words reveal between a strictly defined empiricism (limited to collecting facts, and concluding "no farther than the facts will allow") and speculation from necessarily partial or contingent knowledge informed all geologic science in the late Enlightenment, despite vitriolic claims for pure empiricism advanced by theorists of all camps.[4] The constraints of "present circumstances" might be understood as referring broadly to the impossibility of ascertaining what lay in that realm Martin Rudwick has so appositely termed "Deep Time," referring simultaneously to the buried features of the earth and to its distant past.[5]

3. Franklin was responding to notes Soulavie had sent him of an earlier conversation concerning theories of geologic change: "I wrote it to set him right in some points wherein he had mistaken my meaning" (*Complete Works*, 8:187n).

4. The important European contributions to this debate include Abraham Gottlob Werner, *Kurze Klassification und beschreibung der verschieden Gebirgsarten [Short Classification and Description of the Various Rocks]* (1787); James Hutton, *Theory of the Earth, with Proofs and Illustrations* (1795); Richard Kirwan, *Geological Essays* (1799); John Playfair, *Illustrations of the Huttonian Theory* (1802); Robert Jameson, *Elements of Geognosy* (1804–8); and Robert Bakewell, *An Introduction to Geology, Illustrative of the General Structure of the Earth* (1813). American texts include Parker Cleaveland, *Elementary Treatise on Mineralogy and Geology* (1816); William Maclure, *Observations on the Geology of the United States* (1817); Samuel Latham Mitchell, *Observations on the Geology of North America* (1818); and Amos Eaton, *Index to the Geology of the Northern States* (1818; revised ed., 1820).

5. See Martin J. S. Rudwick, *Scenes from Deep Time: Early Pictorial Representations of the Prehistoric World* (Chicago and London, 1992).

Almost four decades after this exchange of letters, Samuel Seymour was designated official landscape painter for the first scientific exploration of the central plains and Rockies, the Stephen H. Long Expedition (1819–20).[6] Although geological debate had grown more sophisticated and interpretive paradigms were better articulated, methodological ambivalence about the role of speculation persisted. Fieldwork was understood as fundamental to the pursuit of geologic understanding; while speculation, though anathema to some, was widely acknowledged to be equally essential, both the stimulus to research and its interpretative consequence.[7] Seymour's dual commission to seek out landscape subjects of interest to his scientific colleagues yet notable for their "grandeur" thus corresponded to an epistemologic ambivalence informing these colleagues' own thinking.[8]

Although acknowledged to be the first white artist to represent the landforms and geologic formations of the American West, Seymour has not yet received serious study, nor have his "views" been adequately appreciated by scholars for their complexity as both scientific illustration and artistic statement.[9] Seymour has not been recognized as a painter of *geologic* landscapes, even though an important subset of his surviving watercolors from the Long

6. The expedition was known after its commanding officer, Stephen Harriman Long (1784–1864), a major in the newly reorganized Army Corps of Topographical Engineers; see Roger L. Nichols and Patrick L. Halley, *Stephen Long and American Frontier Exploration* (Newark, Del., 1980).

7. For a valuable contemporary appraisal of the state of this debate, see Walter Channing's detailed review of Georges Cuvier's *Essay on the Theory of the Earth* and J. Freeman Dana and Samuel L. Dana's *Outlines of the Mineralogy and Geology of Boston and Its Vicinity* [1818] in *North American Review* 8 (1819): 396–414. John C. Greene, in *American Science in the Age of Jefferson* (Ames, Iowa, 1984), traces the positions taken in this debate back to geological circles in Edinburgh; see also George P. Merrill's still useful *The First One Hundred Years of American Geology* (New Haven, Conn., 1924).

8. "Mr. Seymour as Painter for the Expedition" was to "furnish sketches of Landscapes whenever we meet with any distinguished for their beauty or grandeur—he will also paint miniature likenesses or Portraits if required of distinguished Indians and exhibit groups of savages engaged in celebrating their festivals or sitting in council and in general illustrate any subject that may be deemed appropriate in his art"; Stephen Long to John C. Calhoun, 20 April 1819, "Letters Received by the Secretary of War"; reprinted in *The Papers of John C. Calhoun*, 19 vols. (Columbia, S.C., 1959–90), 4:33–34. I have preferred this version of Long's "Instructions to His Command" to that published later in the *Account* (with most capitals removed and punctuation added) as closer in tone to the instructions Seymour would have received in the field. For the later version, see Edwin James, *Account of an Expedition from Pittsburgh to the Rocky Mountains*, 2 vols. plus an atlas (Philadelphia, 1822–23), 1:3; 14:42 (hereafter *Account*). The Philadelphia edition, substantially emended, was reissued in three volumes in London in 1823, without the atlas. I quote from the original edition (except where indicated), because it was intended for an American audience and better supports the cultural interpretations I propose, but I give corresponding page references from the more readily available reprint of the London edition, vols. 14–17 of *Early Western Travels, 1748–1846*, ed. Reuben Gold Thwaites (Cleveland, 1905).

9. For the most exhaustive treatment given his work to date, see Patricia Trenton and Peter Hassrick, *The Rocky Mountains: A Vision for Artists in the Nineteenth Century* (Norman, Okla., 1983), 20–30.

Expedition (along with engravings after others now lost) surely falls into this category. Indeed, all expeditionary imagery executed in the years of the early republic has been treated in a singularly reductive manner—despite a considerable body of literature demonstrating the importance of *not* mistaking landscape imagery in general for simple transcriptions of nature.[10] My intent in this essay is to establish, in Seymour's expeditionary work, the simultaneous currency of two radically divergent pictorial impulses: mimetic realism on the one hand, which subordinates artistic invention to the facticity of what Franklin termed "actual observation"; and picturesque practice on the other, which draws the viewer in through an idealized reconfiguration of the world, and gives, in Franklin's formulation, "a loose to imagination." While Seymour managed to integrate these contrasting and competing styles of vision within the scope of single compositions by purely gestural means, connecting near to far by *figures looking*, it was his insight into the essential similarities between his art and the science he was representing that provided him the means to unify his compositions not just aesthetically but also thematically.

There has been insufficient attention paid this coincidence in the dialectics animating scientific and artistic "fieldwork" in the period. Just as early American landscape painting was characterized by an ultimately fertile opposition between view and vision, mimesis and imagination,[11] so might the science of the day, earth science in particular, be characterized by a similar opposition between two competing modes of understanding.[12] Late Enlightenment science and early Romantic art, beyond their shared interest in *plein air* investigation as a "method of philosophizing,"[13] were involved in a wide-ranging cultural

10. Key texts in a growing revisionist literature on British and American landscape of the late eighteenth and the nineteenth centuries include John Barrell, *The Dark Side of the Landscape: The Rural Poor in English Painting, 1730–1840* (New York, 1980); David H. Solkin, *Richard Wilson: The Landscape of Reaction* (London, 1982); Bryan Jay Wolf, *Romantic Re-Vision: Culture and Consciousness in Nineteenth-Century American Painting and Literature* (Chicago and London, 1982); Ann Bermingham, *Landscape and Ideology: The English Rustic Tradition, 1740–1860* (Berkeley and Los Angeles, 1986); Angela Miller, *The Empire of the Eye: Landscape Representation and American Cultural Politics, 1825–1875* (Ithaca, N.Y., 1993); and Alan Wallach, "Thomas Cole: Landscape and the Course of American Empire," in William H. Truettner and Alan Wallach, eds., *Thomas Cole: Landscape into History*, exhibition catalogue, National Museum of American Art (Philadelphia, 1994), 23–111.

11. See Edward J. Nygren, *Views and Visions: American Landscape before 1830*, exhibition catalogue, Corcoran Gallery of Art (Washington, D.C., 1986), 3–81. For a survey of early American landscape, see Karol Ann Lawson, "A New World of Gladness and Exertion: Images of the North American Landscape in Maps, Portraits, and Serial Prints before 1820" (Ph.D. diss., University of Virginia, 1988).

12. See Mott T. Greene, *Geology in the Nineteenth Century: Changing Views of a Changing World* (Ithaca, N.Y., 1982), esp. chap. 1, "Hutton and Werner: First Principles."

13. *Plein air* landscape drawing and scientific collecting were contemporary and in many respects analogous late-eighteenth and early-nineteenth-century innovations.

debate regarding the general and the particular—a question of equal relevance whether one was classifying or representing a given geological formation. Although these cultural practices are often seen as having little in common beyond their objects of scrutiny, some critics have followed Rudwick in describing the relationship of geologic art to geologic science in this period as "broadly *complementary*."[14] This conclusion is not surprising once we recognize that both art and science are epistemologies, systems or strategies of knowing, and as such concerned first and foremost with harnessing, regulating, and defining the inquiring gaze.

It now seems commonplace to point out that science by the early decades of the nineteenth century had achieved a new (what we might term modern) self-consciousness concerning both the nature and the limitations of human perception. At the same time, American artists were evidencing, and thematizing in their work, a newly self-conscious awareness and theoretical concern with how the eye moves through space, making "sense of the world" in an active rather than a passive manner. In part, I seek to reveal the contest between styles of seeing that underlies the apparent harmony of Seymour's compositions, and to this end I examine the implications of representational formulas derived from the British picturesque tradition that Seymour applied to the western landscape; and I call attention to the way that Seymour's disposition of forms (especially the relation of figures to the landscapes in which they appear) serves to instruct his viewers how to see, hence how to know.

One of the principal means Seymour devised to unify his geologic landscapes or "scenographic geologic illustrations"[15] both compositionally and thematically was his artistic focus on *origination*, a topos with relevance for a range of intellectual endeavors and institutions including the nation's oldest and most prestigious organizations for the advancement of knowledge, the American Philosophical Society and the Academy of Natural Sciences, both of Philadelphia. At the request of the War Department, these organizations

14. Martin J. S. Rudwick, "The Emergence of a Visual Language for Geological Science, 1760–1840," *History of Science* 14 (1976): 182; see also Rachel Laudan, "The History of Geology, 1780–1840," in Robert Cecil Ottby et al., eds., *Companion to the History of Modern Science* (London, 1989), 322. For an opposing view, see, for example, Ellwood C. Parry III, "Acts of God, Acts of Man: Geological Ideas and the Imaginary Landscapes of Thomas Cole," in Cecil J. Schneer, ed., *Two Hundred Years of Geology in America: Proceedings of the New Hampshire Bicentennial Conference on the History of Geology* (Hanover, 1979), 55.

15. The term "scenographic geology" was coined by Edward Hitchcock in his *Final Report on the Geology of Massachusetts* (Northampton, 1841), to describe the use of scenery in figuring geological formations; see Virginia Lee Wagner, "The Idea of Geology in American Landscape Painting, 1825–1875" (Ph.D. diss., University of Delaware, 1987), 190.

had assisted in defining the Long Expedition's scientific mission.[16] A virtual obsession with questions of origin may fairly be said to typify the scholarly engagements of the age, from political debate fueled by the novelty of American institutions to ethnological debate concerning the prehistory of human cultures to a vigorous natural historical debate on subjects ranging from biological nomenclature to the early history of the earth itself.[17]

It was part and parcel of this widely ramifying interest that questions of origins should have figured into the ambitions of virtually all of Seymour's expeditionary colleagues. Both Long's diplomacy, with its roots in a symbolic language and nationalist rhetoric borrowed from Lewis and Clark, and the ethnographic studies of Thomas Say were driven by interest in "natural" origins.[18] Edwin James's fascination with geognosic origins, meanwhile, informed the field investigations he carried out as the expedition's geologist.[19] Figures of source and origin, while they provided Seymour iconographic means to fulfill his instructions as "Painter for the Expedition," furnished him a pictorial language well suited to his own preoccupation with the issue of origination as it concerned the genesis of his art. Seymour's inflection of the already highly self-conscious idiom of picturesque landscape to thematize, and in some sense mystify, the sources of his own artistic production as *original*—as not subordinate to things as seen—suggests much about the proto-Romantic modernism of American vernacular painting (including scientific illustration) in the early decades of the new republic.

16.   On the role played in western exploration by the American Philosophical Society and the Academy of Natural Sciences (all of the expedition's "scientific gentlemen," including Long, belonged to one or both), see William H. Goetzmann, *Exploration and Empire: The Explorer and the Scientist in the Winning of the American West* (New York, 1966), 182–84; see also Simon Baatz, "Philadelphia Patronage: The Institutional Structure of Natural History in the New Republic, 1800–1833," *Journal of the Early American Republic* 8 (summer 1988): 111–38.
17.   Barbara Stafford, in "Toward Romantic Landscape Perception: Illustrated Travel Accounts and the Rise of 'Singularity' As an Aesthetic Category," *Art Quarterly*, n.s., 1 (autumn 1977), has described natural history as, in fact, *embodying* the "quest for *origins*" (p. 100).
18.   Thomas Say (1787–1834), though officially the expedition's zoologist, was also charged with studying the manners and the customs of Native Americans encountered; see Patricia Tyson Stroud, *Thomas Say: New World Naturalist* (Philadelphia, 1992), chaps. 5–7; and the valuable earlier study by Harry B. Weiss and Grace M. Ziegler, *Thomas Say: Early American Naturalist* (Springfield, Ill., 1931).
19.   Edwin James (1797–1861) was also physician to the expedition. Attention to geologic origins pervades the *Account* (with references to works by Hutton, Werner, Eaton, Maclure, Volney, and Kirwan), climaxing with the arrival at the base of the mountains in June and July of 1820 (see text, below), and after the expedition James pursued the topic in a series of professional papers and articles. The most useful biographical assessment remains Maxine Benson, "Edwin James: Scientist, Linguist, Humanitarian" (Ph.D. diss., University of Colorado, 1968).

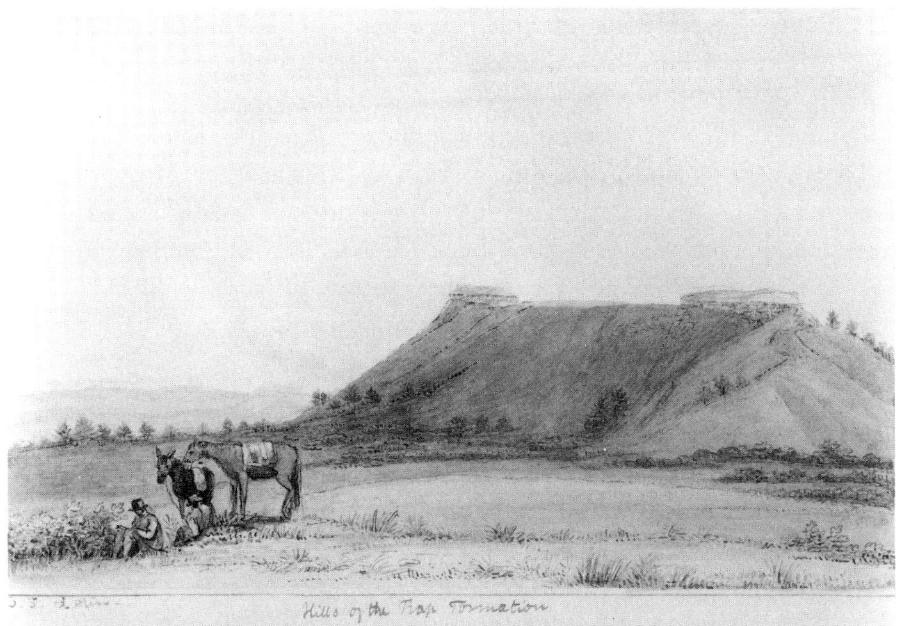

Figure 51. *Hills of the [Floetz] Trap Formation* (1822); watercolor over graphite on paper, 14.5 x 21 cm. Beinecke Rare Book and Manuscript Library, Yale University.

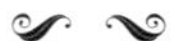

Seymour made the preparatory drawing for his *Hills of the [Floetz] Trap Formation* (fig. 51) in late July 1820, not far from the Arkansas River along the front range of the Rocky Mountains. We see in the middle distance an unusual formation of a kind then referred to by some geologists as floetz rock, a German mining term for formations with predominant horizontal stratification. This particular occurrence is described at length in the Long Expedition's official narrative report, of which James was the nominal compiler and principal author, as a "symmetric disposition of truncated cones, sometimes insulated, sometimes grouped, and by elevated plains, both extremities of which are crowned by a conical rising."[20]

Seymour has placed the figures of two members of the party (and their mules) in the lower left foreground of the image. The seated figure nearest us,

20. *Account*, 2:404; 17:210. The geological report in the appendix to the *Account*, from which this passage is quoted, has previously been credited either to Long (to whom it was submitted) or to Augustus Jessup (the original expedition geologist, listed in the table of contents of the British editions as its author), but is certainly the work of James. The phrase "elevated plains, both extremities of which are crowned by a conical rising" is in italics in the original.

with his knees up and his head down, his attention focused on his open note-book, may be read equally well as a geologist taking field notes or as an artist working in his sketchbook—in the context of the expedition, as either James or Seymour himself. That this illustration should chronicle the geologic field-work whose results it also represents—James studying the "symmetric disposi-tion of truncated cones," then writing the description we have read—may seem tautological. But so too does the visual evidence—the presence of Sey-mour's signature immediately below—that what we see is a self-portrait of the artist in the process of capturing this very image. The anomaly in both cases is the same, of course: that this figure, whether scientist or artist, has his back turned to the world that he transcribes—if, indeed, transcription this is. The geologic description memorialized *in* the watercolor, like the watercolor itself, is less an act of representation than one of self-reflection, a text (judging from the body language of its maker) essentially closed upon itself, or open upon the noumenal world in only a highly mediated fashion. The epistemologic value of this passage is if anything reinforced by its conventional position in Seymour's composition as *repoussoir,* as a framing device informing our own relationship as viewers to the world we are invited to behold.

As we can see, comparing the watercolor to James's description, Seymour took the work of geologic illustration very seriously, but only as "pre-text" for his own self-conscious affirmation of the role played by imagination—not just in the work of scientific illustration but in science itself. For if what we have here is in one sense a narrative of fieldwork (whether *plein air* sketching or geo-logical research) understood in Franklin's terms as a method that "proceeds upon actual observation, makes a collection of facts, and concludes no farther than those facts will warrant," in another sense Seymour's *Hills of the [Floetz] Trap Formation* offers a competing vision—science as the work of reverie. It would be difficult to exaggerate the significance of this defeat of positivist expectation on Seymour's part, the more so given that his recognition was shared. In a paper read before the American Philosophical Society—the one on which the strictly mineralogical treatment quoted above was based—James himself acknowledged that the "appearance and position" of such curious for-mations "lead almost involuntarily those least attached to visionary theories into speculations concerning their origin."[21]

In this view, taken to its extreme, geology is but a science of representation with its analogue in art; and the work of the geologist, like that of the artist,

---

21. Edwin James, "Remarks on the Sandstone and Floetz Trap Formations of the Western Part of the Valley of the Mississippi," *Transactions of the American Philosophical Society,* n.s., 2 (1825): 209. This paper was written in April 1821 and read before the Society on 17 August of that year, at roughly the same time that Seymour was coloring his sketch.

involves inscription of the world with meaning as much as transcription of meanings pre-inscribed in the world. Even the term "Trap Formation" in the watercolor's title calls punning attention to the circumstance of representation closed upon itself. Transforming static geologic illustration into a mode of history painting writ small, Seymour thematized both the work of science and the work of art as concurrently descriptive and inscriptive. It is not that either serves as a vehicle for the other, but rather that the two interdepend.[22]

This interdependence was certainly reflected in the various narratives produced by James, who was responsible for most of the contemporary written reports of the expedition's findings. James describes the environs of the floetz rock formation visible in Seymour's watercolor as "abound[ing] in scenery of a grand and most interesting character," going on to comment effusively:

> The contrast of colours which is here seen, often produces the most brilliant and grateful effects. The deep green of the small procumbent cedars and junipers, with the less intense colours of various kinds of deciduous foliage, acquire new beauty by being placed as a margin to the glowing red and yellow which is seen on the surface of many of the rocks. In the narrow but verdant valleys, small pyramids and columns of the purest white are met with, standing solitary and detached from any surrounding rocks.[23]

Here James the geologist speaks the language of art, employing a palette that closely resembles Seymour's own. He describes the sandstone out of which the "conical risings" emerge as "var[ying] in colour from vermilion red to dark brown, and sometimes to various shades of yellow and grey." Conversely, Seymour's palette might be viewed as a coloristic analogue to the Geological Alphabet proposed by Amos Eaton, who had instructed James in earth science at Middlebury College, "consisting of nine simple minerals, an acquaintance with which will enable any one to spell out any rock with facility."[24]

22. Rudwick, in "A Visual Language for Geological Science," first argued that geologic illustrations in this period constitute a "visual mode" of scientific communication.

23 James, "Remarks on the Sandstone and Floetz Trap Formations," 197–98.

24. John White Webster, review of the revised edition of Eaton's *Index to the Geology of the Northern States*, in *North American Review* 11 (October 1820): 231–32. Seymour's palette, mixed from commercially produced cakes of "Brown Madder, Scarlet Lake, Red Lake, Bronze, Virmillon, B[urn]t Umber, B[urn]t Sienna, Yellow Oker [Ocher], Bister, and Gamboge," along with a variety of other pigments, featured the same range of colors James employs in his descriptions; see "Exploratory Expedition under Major Stephen Long," 3d Auditor's Accounts, Department of the Treasury, National Archives, purchase vouchers 87 (9 March 1819) and 48 (17 March 1820). For fuller transcriptions, see Kenneth Haltman, "Figures in a Western Landscape: Reading the Art of Titian Ramsay Peale from the Long Expedition to the Rocky Mountains, 1819–1820" (Ph.D. diss., Yale University, 1992), appendix D, 267–70.

A similarly rich combination of scientific and aesthetic interest can be seen throughout the *Account*. Where James portrays the scenery along the banks of the Ohio as "eminently beautiful" but "deficient in grandeur and variety," he relies on the conventional vocabulary of the picturesque, familiar from travel narratives of the period as well as aesthetic treatises. He paints the scene as though it were a picture: "The hills usually approach on both sides, nearly to the brink of the river; they have a rounded and graceful form, and are so grouped as to produce a pleasing effect." Such chiastic prose echoes the form of the landscape it describes. With a change of voice but without pause, James goes on to note, now employing the vocabulary of a field geologist: "[i]n a few instances near the summits of the hills, the forest trees become so scattered, as to disclose here and there a rude mass, or a perpendicular precipice of gray sandstone, or compact limestone, the prevailing rocks in all this region."[25]

Just as Seymour in his landscapes advanced the expedition's scientific agenda where choice of subject was concerned, the science that he represented based some of its reading of the natural world on unabashedly aesthetic categories of analysis. Thus Seymour's figure of an artist-scientist-explorer looking inward to describe the world was perhaps not so unscientific as it might at first appear. As Marcia Pointon has argued, what artists like Seymour and field geologists like James shared was a language of picturesque description. Neither a strictly topographic mode of pictorial representation nor the language of rigorously objective empirical description sufficed to convey the manifest evidence of geologic history. A more "painterly technique" was called for.[26]

Hills of the [Floetz] Trap Formation is in this respect a watercolor image quite typical of geologic landscapes of its era. Color and wash were added over penciled drawings on a sheet removed by Seymour from his sketchbook, a process apparently designed to preserve the immediacy of his original impressions. This transition from graphite field sketch to finished composition recapitulates the larger movement in the English-derived watercolor tradition beginning around 1760 from a rather linear military topographicism to a land-

---

25. *Account*, 2:19; 14:62. On the connection between pictorial and rhetorical conventions of the picturesque, see Christopher Hussey, *The Picturesque: Studies in a Point of View* (London, 1927). For an overview of the interdependence of art and science in Philadelphia in this period (where James composed the *Account* and Seymour his finished images), see Edgar P. Richardson, "The Athens of America, 1800–1825," in Russell F. Weigley, ed., *Philadelphia: A Three-Hundred-Year History* (New York, 1982), 241–51.

26. See Marcia Pointon, "Geology and Landscape Painting in Nineteenth-Century England," in L. J. Jordanova and Roy S. Porter, eds., *Images of the Earth: Essays in the History of the Environmental Sciences*, British Society for the History of Science, monograph 1 (Cambridge, 1979), 86.

scape idiom concerned with affect and emotion, and with higher meaning—
that is, the picturesque.[27]

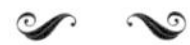

At the time he was named to the Long Expedition in 1819, Seymour enjoyed
a local reputation in Philadelphia as a landscape painter on the margins of his
regular trade as a commercial engraver. Born in England, or possibly Scotland,
Seymour came to Philadelphia as a young man; by 1795 he had completed
five plates for Thomas Dobson's American edition of the *Encyclopaedia
Britannica*. Subsequently he found employment in the shop of the influential British-born engravers and artists William and Thomas Birch, executing
some compositions of his own design but chiefly producing popular prints
after paintings by others.[28] Though Seymour was by no means prominent—
in fact there is some evidence that, having reached middle age without making his professional mark, he viewed employment on the Long Expedition as
an opportunity for advancement—he was not altogether unknown.[29] As an
associate member of the Society of Artists, Seymour had shown in annual
exhibitions at the Pennsylvania Academy alongside the leading artists of his
day, including William Rush, Thomas Sully, James and Rembrandt Peale,
Alvan Fisher, Robert Salmon, Thomas Doughty, John Wesley Jarvis, and
Thomas Birch, although none of his works from this period has survived.[30]

We can guess, however, from both the titles of works that he exhibited and
his later expeditionary watercolors, that the pictorial conventions of the British
picturesque, a style of representation with great commercial and artistic currency in early republican Philadelphia, significantly informed his vision.

27. See Bruce Robertson, "Venit, Vidit, Depinxit: The Military Artist in America," in Nygren, *Views and Visions*,
     83–104. William Gilpin (*Three Essays: On Picturesque Beauty; On Picturesque Travel; and On Sketching
     Landscape; To Which Is Added A Poem, On Landscape Painting* [London, 1792], 62–66) recommended and
     justified this technique in an eighteenth-century British context (see text, below).
28. See Robert D. Arner, *Dobson's* Encyclopaedia: *The Publisher, Text, and Publication of America's First*
     Britannica, *1789–1803* (Philadelphia, 1991), 142, 231–32. Seymour's arrival in the New World has tradi-
     tionally been dated to later in the decade; see, for example, David McNeeley Stauffer, *American Engravers
     Upon Copper and Steel: Biographical Sketches and A Check-List of the Works of the Earliest Engravers*, 2 vols.
     (New York, 1907), 1:244, followed by Trenton and Hassrick, *The Rocky Mountains*, 347 n. 36. Robert D.
     Mussey Jr. has recently uncovered evidence that Seymour may have arrived in America as early as 1784 and
     have received training as an engraver, perhaps in Boston, before moving to Philadelphia (personal correspon-
     dence, 19 May 1997).
29. Joshua Shaw and perhaps John Lewis Krimmel had declined the position before Seymour was approached
     (*Papers of Calhoun*, 3:395–96, 422–23).
30. See Edward J. Nygren, "Art Instruction in Philadelphia, 1795–1845" (M.A. thesis, University of Delaware,
     1969), 3–6, 61, 208–12. Seymour showed landscapes in the Academy exhibitions of 1811 and 1814.

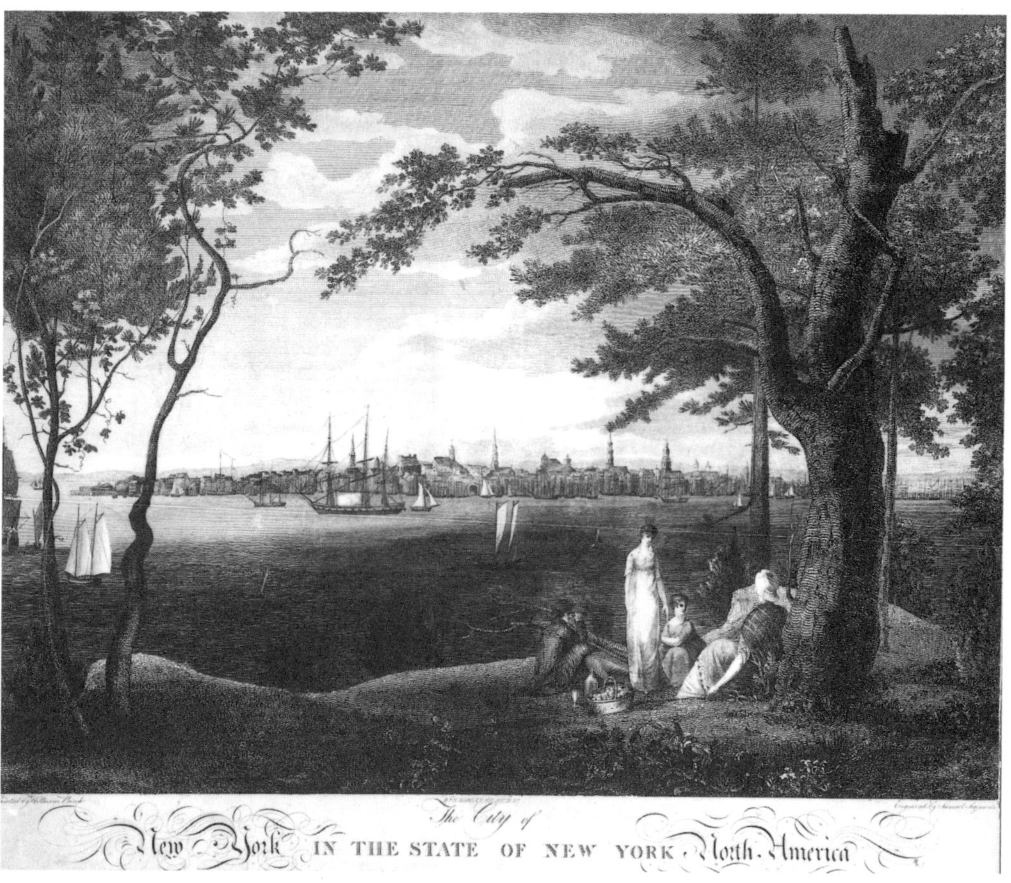

Figure 52. *View of the City of New York in the State of New York from Long Island* (1803); line engraving, 47.8 x 60.8 cm. Huntington Library.

Seymour, for example, copied prints of exotic locales from John Pinkerton's *A General Collection of the Best and Most Interesting Voyages and Travels, In All Parts of the World* for the American reprint edition published in Philadelphia between 1808 and 1813.[31] His own *View of the City of New York* (fig. 52), Manhattan Island seen from from Brooklyn across the northern tip of New York harbor, exemplifies the compositional techniques he had acquired, adopting an Old World pictorial idiom to various New World subjects. The city itself, though the nominal subject of the engraving, is visible only in the distance as a diminutive array of masts and spires, viewed across a broad body of calm water in the midground that separates us from the world we are invited

31. See John Pinkerton, *A General Collection of the Best and Most Interesting Voyages and Travels, In All Parts of the World*, 6 vols. (Philadelphia, 1808–13), vols. 2–5; see figs. 11 and 13, below.

to survey. This spatial interruption is at once naturalized and rationalized—made narrative—by a party of elegantly attired picnickers underneath a pair of graceful trees that frame and center the view of distant wharves and buildings. The human presence in the immediate foreground of the image serves to render the distant city psychologically accessible. But it is the discontinuity that in fact *constitutes* the view, making it visible, reminding us of where we stand in relation to the distance.

Picturesque landscape theory received its classic formulation in the writings of late-eighteenth-century British aestheticians, notably William Gilpin, whose works enjoyed wide circulation in America.[32] Gilpin favored this same basic tripartite structuring of picture space.[33] The titular subject of a composition (whether hills by the base of the Rocky Mountains or the skyline of New York) was typically situated at a substantial distance from the picture plane and separated from the viewer (and so constructed as a visualizable entity) by an expanse of empty space. The subject was introduced and glossed by *repoussoir* figures whose presence serves in some sense to explain our own, and by whose attitudes and gestures our perceptions are informed. In such works a purely visual accessibility replaces actual accessibility, reverie actual penetration on the ground. Or, to state this in slightly different terms, good picturesque landscape makes its principal appeal not to the senses but to the imagination.

In harbor views this was all well and good, serving to infuse a commonplace scene with atmosphere. We can be reasonably certain, however, that Seymour's employers in the War Department—despite the ambiguities in his commission already alluded to—intended that he provide a set of reproducible watercolor images constituting a pictorial record of real access on the ground. In the words of Calhoun, who as secretary of war personally approved the hiring of both Seymour and Titian Peale (the latter responsible for scientific specimen studies): "The object of the Expedition, is to acquire as thorough and accurate knowledge as may be practicable, of a portion of our country . . . as yet imperfectly known."[34] These instructions seem incompatible, in principle if not in practice, with the picturesque talents of an artist like Seymour, whose

32. Such works were a staple of the Philadelphia book trade from the turn of the century onward. The Library Company of Philadelphia held the largest collection of texts of aesthetic theory in the country. See Janice G. Schimmelman, "European Treatises on Art and Essays on Aesthetics Available in America through 1815," *Proceedings of the American Antiquarian Society* 93, pt. 1 (April 1983): 95–195.

33. Gilpin, *Three Essays*, "Notes to the Poem," ii.

34. Instructions to Stephen Long, cited in the "Preliminary Notice" to *Account*, 1:3 (Philadelphia edition only). The record is equivocal, but Calhoun probably agreed to include artists among the exhibitionary command because of a wish to serve the ends of science, linked as these were with territorial expansion; and an intuition of the potential political appeal of "authenticating imagery" in justifying both present and future expenditures on exploration.

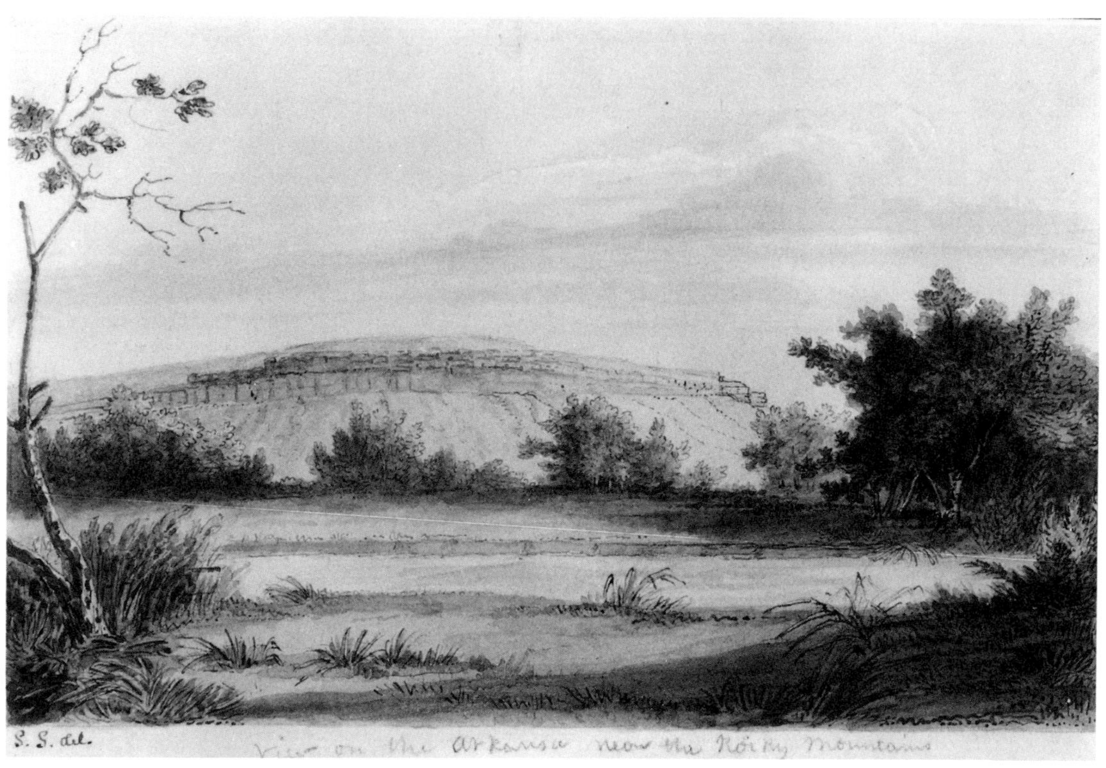

Figure 53. *View on the Arkansa[s] Near the Rocky Mountains* (1822); watercolor over graphite on paper, 13.5 x 20.5 cm. Beinecke Rare Book and Manuscript Library.

skills and ambitions were ill-suited to the transcription of observed reality. Such "telescopic pleasures of the eye," in Gilpin's words, were "very little allied to the pleasures of the painter," who relied instead on compositional devices whose function was to keep the distance at a distance.[35]

Seymour's *View on the Arkansa[s] Near the Rocky Mountains* (fig. 53), based on a field sketch executed in mid-July, suggests a contest between two modes of vision.[36] Our view to the rounded cliff of reddish sandstone in the far midground lies across a body of water, presumably the Arkansas River, which transects the composition laterally. In combination with a line of trees positioned further in the distance, this thin line of river interrupts our possible advance, dividing us from the wall of geologic rock we view. The further containment of the formation by a gracefully arching birch at foreground left and a clump of trees at midground right, along with the mediating water, consti-

35. Gilpin, *Three Essays*, 201.
36. *Account*, 2:42; 16:32.

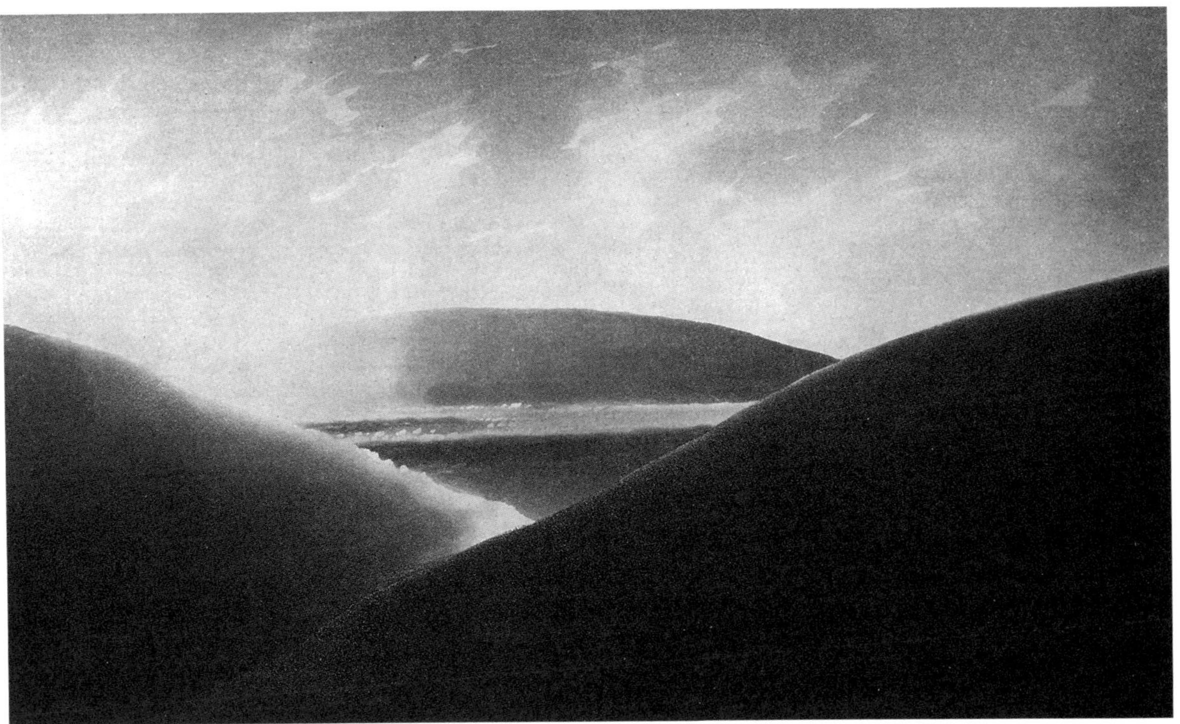

Figure 54. William Gilpin, *Scene without Picturesque Adornment* (1792); etching and aquatint, 13. 6 x 10.9 cm. *Three Essays* (London, 1792), plate 2. Huntington Library.

tutes a "picturesque signature" on Seymour's part that tempts the practiced viewer's gaze from its scientific object—the distant hillside—into meditation and reflection.[37]

It should not surprise us that the ostensibly topographic content of this image takes its pattern not from nature but from art. In his treatise *Three Essays: On Picturesque Beauty; On Picturesque Travel; and On Sketching Landscape*, quoted earlier, Gilpin illustrates the case against overreliance on "the Beautiful" with a pair of exemplary contrasting images of the same scene, first without and then with "picturesque adornment." It is easy to recognize in the former (fig. 54) the underlying structure that Seymour, in *View on the Arkansa*, has successfully modified. The eye is engaged by the addition of "frictionalizing" passages of visual

37.   Seymour makes his organizing presence more explicit still, placing his initials below the tall birch that alone spans picture space entirely, as the viewer is invited to do optically and in imagination. Compare these lines from Ralph Waldo Emerson's "Nature" (1836): "There is a property in the horizon which no man has but he whose eye can integrate all the parts, that is, the poet"—cited by W. J. T. Mitchell in "Imperial Landscape," his own essay in the anthology he recently edited, *Landscape and Power* (Chicago, 1994), 14.

interest, and yet, for the sake of aesthetic appeal, regular symmetry has not been sacrificed entirely. This solution essentially conforms to Gilpin's recommendation: the eye's passage is neither unimpeded nor impeded absolutely; but the viewer's principal (or at least most immediate) reward is the discovery of an interesting passage through picture space.[38]

The picturesque challenge Seymour's landscapes pose the viewer is in every way compatible with his commission, as it involves the viewer in the quiet drama of exploration, that of discovering a passage through picture space. In *View Near the Base of the Rocky Mountains* (fig. 55), a work dating to the first week in July 1820, the three small figures we come upon at foreground center, white men in brown expedition uniforms and black hats, bearing rifles, serve in this respect as our surrogates. They stand with their backs to us, gazing out across a grassy midground to a line of jutting rock formations rising in the middle distance.[39] Beyond, dense and dully colored rolling hills extend to the far horizon. In a gap in the low foliage at foreground center, just behind these standing men and at our feet, a pool of pale blue water reflects the soft gray of the sky. In the middle distance, the overriding horizontality of the composition is broken by a pair of roughly centered vertical outcroppings of white basalt—framed by red sandstone bluffs to their left and a rounded, tree-crested hillock to their right—that echo the men's figures at a considerable distance from the picture plane, anticipating human entry into picture space. The pointing gesture of the figure farthest to the right, most likely James, makes literal this formal *repoussoir*. Our gaze passes, as it were, over these three men's shoulders; we look where they look, imagine ourselves going where they seem to wish to go.

But where? No obvious destination presents itself. We may well be "near the base of the Rocky Mountains," as the title indicates, scripted in the artist's hand beneath the image; but there are no "mountains" in view. There are any number of possible passages through this landscape, no one of them clearly preferable to the rest. As viewers we share a dilemma with the small figures who precede us: Which way to turn? The opposing curves of the enlisted men's upraised rifles on the left and James's pointing gesture on the right contribute to this lack of visual resolution. Together they form a figure of ambivalence, of hesitation. The drama of the image—a purely *visual* drama, picturesque in its essence—lies precisely here, in this uncertainty about location and direction.

38. Gilpin, *Three Essays*, 19.
39. See *Account*, 1:500; 15:279.

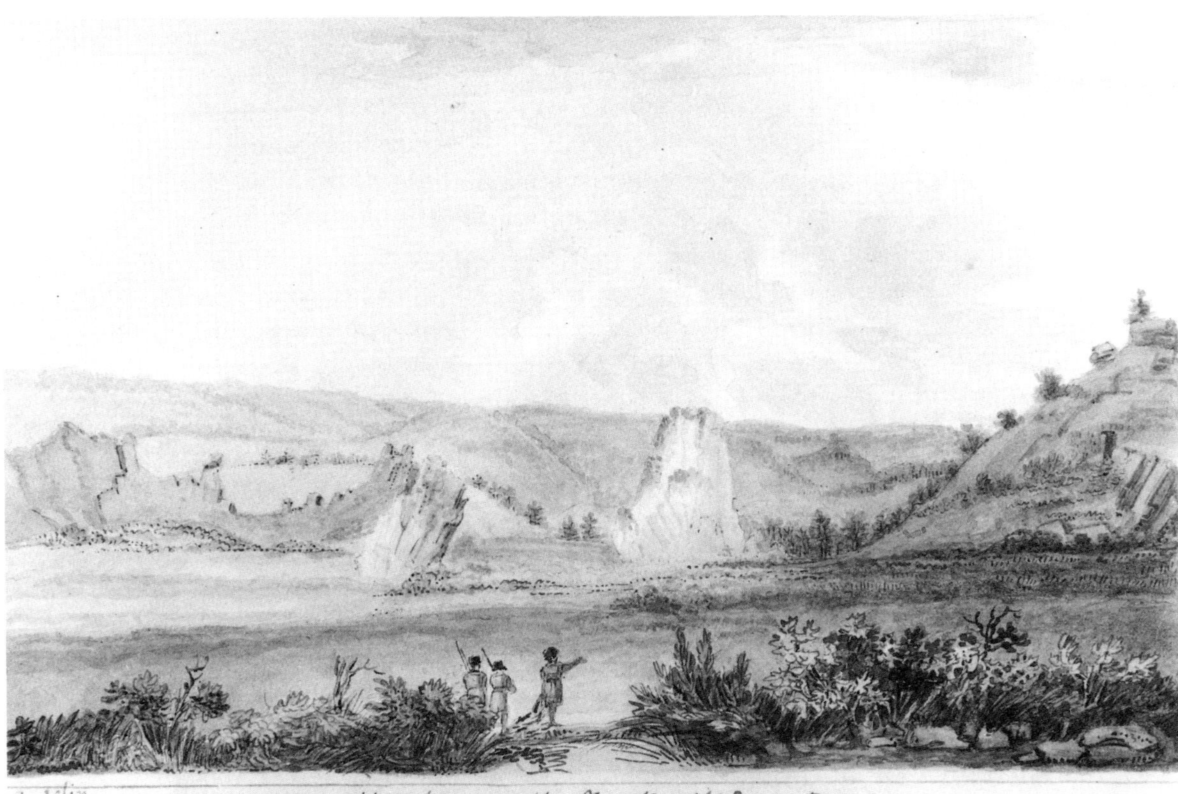

Figure 55. *View Near the Base of the Rocky Mountains* (1822); watercolor over graphite on paper, 14.5 x 21 cm. Beinecke Rare Book and Manuscript Library.

The body of blue water at our feet, exactly on the axis of our entry into picture space, like the waters of the East River in *View of the City of New York*, reinforces our separation from any but a visual sharing in the journey ahead. And indeed, if the Rocky Mountains can be said to appear anywhere in Seymour's composition, it is at our feet, in the words disingenuously scripted by the artist at the lower margin of his sheet, between ourselves and our explorer-surrogates and just adjacent to this pool of water—that is, not the "real" Rocky Mountains but those made available to us via his vision and our own reflection.

Uncertainty about what lies off in the distance is, in other words, a picturesque technique enabling Seymour the better to engage his viewers in acts of vicarious contemplation. As Gilpin himself explained in "Instructions for Examining Landscape," his most frequently reprinted essay: "When we see a pleasing scene, we cannot help supposing, there are other beautiful appendages connected with it, tho' concealed from our view." The artist merely exploits this tendency, seeking to "interest the imagination of the spectator, so as to create in him an idea of some beautiful scenery beyond such a hill, or such a promontory, which intercepts the view." Gilpin describes the deception that results as "like the landscape of a dream."[40] In the context of an expedition seeking "to acquire as thorough and accurate knowledge as may be practicable," however, what this figure takes as its true subject is the very possibility of knowing.[41]

In Seymour's image, James and his men stand in metonymically for us as viewers in our first encounter with an unknown world. We see the geologic features of the western landscape that they have come to study as they study them, sharing in what amounts to a community of longing for a better view than this *View* offers us. For this world has been arranged in such a manner as to render material and cognitive passage through it equally problematic. Ambiguous lines of access and discontinuities cast an air of doubt upon the ultimate adequacy of scientific mensuration to take hold of alien terrain—a Romantic challenge to empiricism, serving here to inscribe a picturesque formula with explicitly epistemologic content. In strictly geologic terms, this image serves to introduce a whole series of Seymour's expeditionary watercolors by opening a panoramic window upon the principal formations discovered "Near the Base of the Rocky Mountains." The grouping of basaltic columns at midground center and sandstone cliff at midground left—formations crucial to the theorizing about the origins of the earth—recur in a series of more detailed images, all dating to about the same period in early July.

The contingency of this small white presence in the vast western landscape corresponds to the uncertainty and frustration recorded by James in his journal and later published in the *Account,* owing in part to the roughness of the terrain, experienced as a discontinuity between visual and actual scale. "From our camp," he wrote, "we had expected to be able to ascend the most distant summits then in sight and return the same evening; but when night overtook us, we found ourselves scarcely arrived at the base of the mountains."[42] He was furthermore aware of what was at stake, professionally as well as scientifically, in these investigations; that a "thorough acquaintance with [the] geological character [of the region] would, in all probability, lead to the most important

40. Gilpin, "Instructions for Examining Landscape," in *Three Essays,* 14–15.
41. See Bryan Jay Wolf's definition of the picturesque as "a mode of epistemology," as "more than a theory of landscape composition, . . . an effort to define the very way in which nature is perceived"; "Revolution in the Landscape: John Trumbull and Picturesque Painting," in *John Trumbull: The Hand and the Spirit of a Painter,* ed. Helen A. Cooper, exhibition catalogue, Yale University Art Gallery (New Haven, Conn., 1982), 207.
42. *Account,* 2:2; 15:287.

conclusions, in forming a correct 'theory of the earth.'"[43] Even as he stood facing the mountains, unsure of what to make of them, John White Webster, writing in the prestigious *North American Review* (and alluding to the expedition then in progress), was reminding his readers: "The eyes of the geologists of Europe are turned to this continent, in the grand and extensive formations of which, they anticipate with no feeble interest the solution of some of the long contested problems of their favourite science."[44]

American interest in geological phenomena was overtly nationalist, as evidenced by the enthusiastic coverage of the expedition's scientific progress in the popular press and echoed by James himself in the *Account*, the "Preliminary Notice" of which reads in part: "We cannot but hope . . . that the time will arrive, when we shall no longer be indebted to the men of foreign countries, for a knowledge of any of the products of our soil, or for our opinions in science."[45] It was with this agenda that Thomas Cooper of the American Philosophical Society had been asked to submit "a schedule of queries in geology and mineralogy" as a program of investigation for the benefit of the expedition geologist.[46] There is ample evidence in the *Account* and elsewhere, including James's own diary and letters, that he understood this geological fieldwork to lie at the crux of the expedition's scientific investigations, and correctly so, if the critical response to his report is any indication. As one reviewer wrote in 1823, "These volumes contain a very valuable body of facts respecting the natural history of the Western Country, and particularly its geology, a subject probably of greater interest than the geology of any other part of the world."[47] Horace Hayden, writing in Baltimore's *Methodist Recorder* the following year, concurred: "[P]erhaps no country upon its surface affords a more suitable field for scientific research, or more ample opportunities, and numerous facts from which to form correct ideas . . . than the continent of North America."[48]

Although James was an anxious man by temperament, plagued with personal doubts exacerbated by worry about money and professional standing,[49] these were not the only or even the principal pressures that he felt.

---

43. Ibid., 2:379; 17:175.

44. *North American Review* 11 (1820): 227.

45. *Account*, 1:2; 14:36.

46. *Papers of Calhoun*, 3:655–56, 692–93, 711; 4:53.

47. *Western Quarterly Reporter of Medical, Surgical, and Natural Science* 2 (1823): 143.

48. Horace H. Hayden, "An Inquiry into Some of the Geological Phenomena To Be Found in Various Parts of America, and Elsewhere," *Methodist Recorder* 1 (1824): 27.

49. See Edwin James, Letterbook (ca. 1819–24), Beinecke Rare Book and Manuscript Library, Yale University, passim.

He complained about his lack of proper access to books and to a good cabinet of minerals, both in the field and at Smithland, a small town in Kentucky where, too ill to travel, he drafted his preliminary report, "Observations on the Mineralogy and Geology of A Part of the United States West of the Mississippi," completed in January 1821. Portions of his introduction to this report were published only in the British edition of the *Account,* where he apologizes for a lack of theoretical sophistication, owed to "the unsettled and progressive condition of geognostic science" in general as well as to the particular circumstances in which he was writing.[50] The problem of interpreting ambiguous or conflicting evidence of geologic origins preoccupied James throughout the course of the expedition but did so particularly in July, when he was in closest proximity to the mountains, frustrated in his efforts to "philosophize" even from facts gathered through his own fieldwork. His instruments were inadequate; ironically, he fretted that his exercise in "actual observation" had put him out of touch with the "whims and notions" of contemporary geology.[51] James also complained about how little opportunity he had been allowed by Long to study formations like those spread across the distant midground of *View Near the Base of the Rocky Mountains.* He felt that he simply needed more time to evaluate this evidence of the processes by which the earth had been formed—the heart of the most pressing geological debate of his generation, and the reason for the heightened international interest in his findings.

Neptunian theory held that the surface of the earth had once been submerged entirely beneath a vast primeval ocean, out of which mineral deposits had precipitated and later been exposed when the waters receded. In the words of Abraham Gottlob Werner: "Our earth is a child of time and has been built up gradually"—that is, by sedimentation modified by gradual erosion.[52] The "steep inclinations" we see James confronting in *View Near the Base of the Rocky Mountains* were but unusually dramatic instances of a class of geologic anomaly recognized for decades to pose the most decisive challenge to the Wernerian paradigm. Two explanations were offered to shore up this view, attributing these phenomena either to residual conformity to hypothetical foliations in the primitive rock or to slippages in the deposited precipitate, neither explanation

---

50. See *Account,* 17:183.
51. Edwin James to Amos Eaton, 8 December 1820, Gratz Collection, Historical Society of Pennsylvania.
52. Abraham Gottlob Werner, "Allgemeine Betrachtungen über die festen Erdkorper" [1817]; cited in translation in Frank Dawson Adams, *The Birth and Development of the Geological Sciences* (Baltimore, 1938), 220–21. Werner is considered responsible for the modern scientific formulation of Neptunian theory.

entirely satisfactory. Thus there emerged around the turn of the nineteenth century a competing paradigm known in its day as Plutonian or Vulcanian theory, which proposed that such anomalies were the result of violent upheavals caused by subterranean fire. This case was first put forth in a cogent fashion by James Hutton, who agreed with Werner that most geologic formations were best described as sedimentary but was convinced nonetheless that only the effects of violent fracture and dislocation could account for the available evidence.

What the American West represented was the promise of decisive *new* evidence. James, obliged to take an interpretive stand in terms of this debate, lamented to friend and mentor Amos Eaton (whose own writings reveal both a debt to Werner's system of classification and the current skepticism toward his geognostic theories): "There are several extensive rocky formations which I have no name for and numerous facts so widely different from what I expected to find that I am at a loss to know how to talk about them."[53] That summer Eaton had sent James page proofs of the revised edition of his own *Index to the Geology of the Northern States,* and James hastened to inform him that he had found the mountains themselves to be composed of primitive rock, a blow to the Wernerian position.[54] But even this key discovery raised more questions than it answered. As James put it in the *Account,* although the "highly primitive character" of the Rocky Mountains did appear to prove Werner wrong, the evidence in support of the "igneous hypothesis" advanced by Hutton remained inconclusive.[55] Furthermore, as James readily admitted to Eaton, "it was not in the mountains that my ignorance was most confounded." It was the terrain immediately subjacent to the mountains, interrupted by the very "range of naked perpendicular and lofty rocks" pictured by Seymour, that "presented the greatest variety of unexpected appearances, and facts which I have not ingenuity enough to reconcile with received opinions."[56]

53. Edwin James to Amos Eaton, 8 December 1820.
54. Eaton, who considered the composition of the mountains crucial to the ongoing debate, had written the previous March to Henry Rowe Schoolcraft, another former student, who had recently returned from a geologic survey of the Lake Superior copper region: "Have you any knowledge of the strata constituting Rocky Mountains? Is it primitive, or is it graywacke like Catskill Mountains? I have said, in a note, that, after you and Dr. E. James set foot upon it, we shall no longer be ignorant of it." See Henry Rowe Schoolcraft, *Personal Memoir of a Residence of Thirty Years With the Indian Tribes on the American Frontiers* (Philadelphia, 1851), 48. There is evidence that James too was reading Schoolcraft's report on the region's geology as he wrote his own. See Edwin James, "Notes of a part of the Expd. of Discovery Commanded by S. H. Long Maj. U.S. Eng. &c &c" (hereafter *Notes*), 3 vols. bound as one, Special Collections, Butler Library, Columbia University, 2:81–83, entry for 12 July 1821.
55. *Account,* 2:394n; 17:200n.
56. Edwin James to Amos Eaton, 8 December 1820.

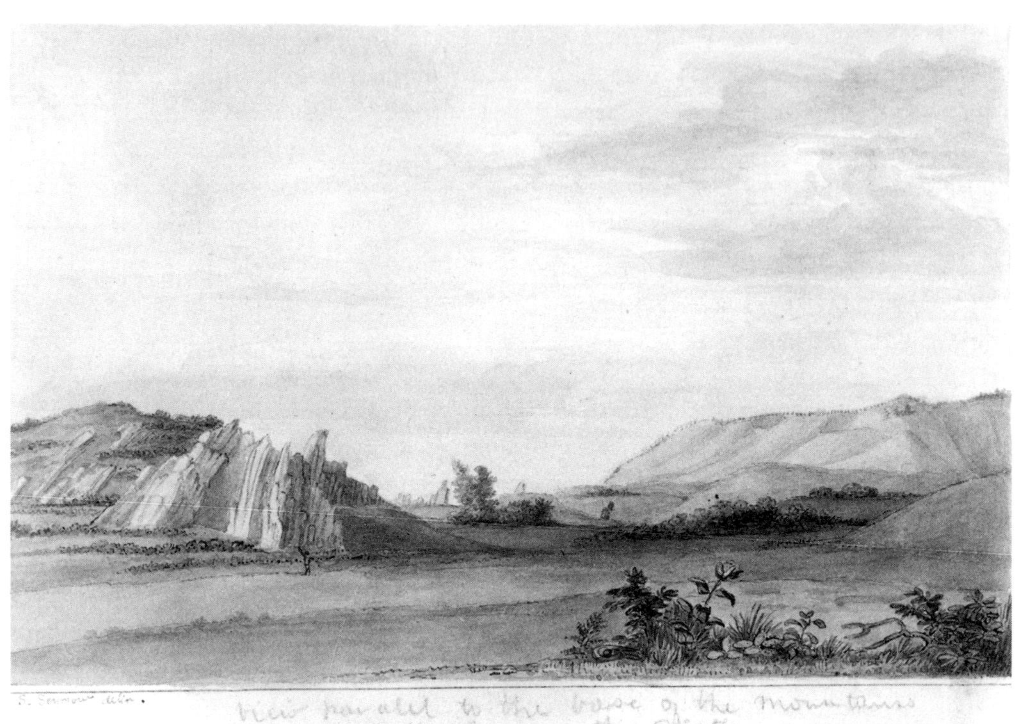

Figure 56. *View Parallel to the Base of the Rocky Mountains at the Head of the Platte* (1822); watercolor over graphite on paper 14.6 x 21 cm. Beinecke Rare Book and Manuscript Library.

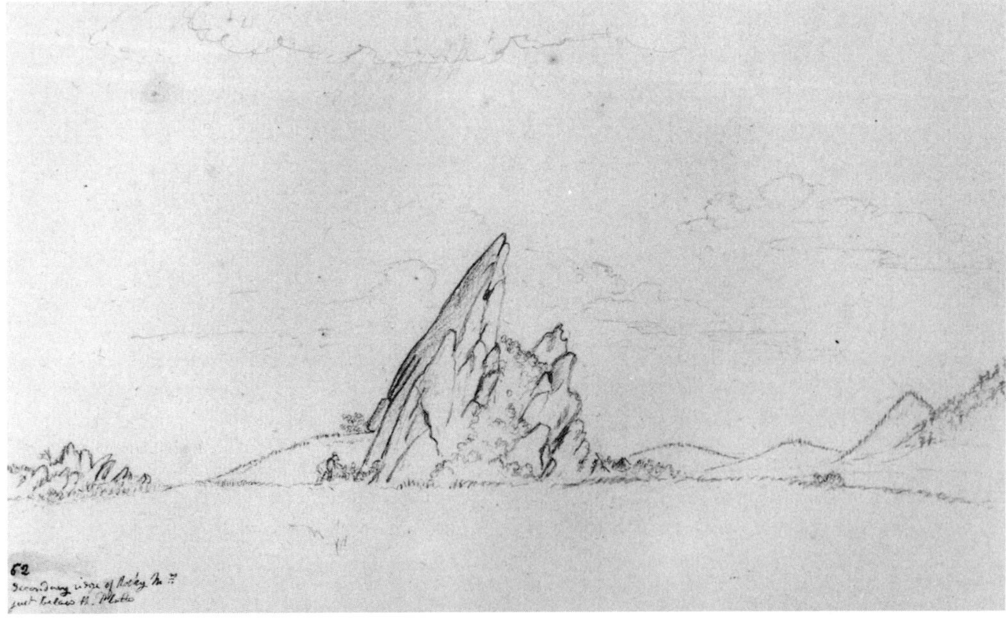

Figure 57. Titian Ramsay Peale, *Secondary Ridge of the Rocky Mountains* (5–6 July 1820); ink over graphite on paper, 13.2 x 21.3 cm. Yale University Art Gallery (IV, 7r).

Seymour's *View Parallel to the Base of the Rocky Mountains at the Head of the Platte* (fig. 56) focuses closer attention on the same cluster of white basaltic columns that appear in the distant center midground of his *View Near the Base of the Rocky Mountains.* As the columns evidence the steepest inclination, they attract and compel the greatest geologic interest. There is a human presence in this image as well, but one far more diminutive—a solitary soldier standing at attention, shouldering a rifle. Almost lost to view near the foot of the formation, he symbolizes not human order and control so much as the will to human order and control.

Titian Peale, the expedition's scientific illustrator, was apparently intrigued by this formation as well, for he executed a careful study of it from the identical vantage and on the same occasion, *Secondary Ridge of the Rocky Mts. just below the Platte* (fig. 57), a work that has just recently come to light.[57] Here again the nearly invisible figure of this soldier is positioned in the midground, just to the left of the composition's center, at once dwarfed by the enormous rock formation before which he stands and echoing its form. Peale's sketch is principally the study of a geologic curiosity appropriately centered in his composition, his small soldier primarily a staffage element. Seymour's watercolor, read in the epistemologic context established by related *Views,* infuses his portrait of this same "singular" formation with a suggestion of doubt regarding the ability of soldier-scientists to make sense of the geologic evidence it represents. In both of these treatments, the physical universe is ordered around symbols of applied intelligence—guns on the one hand, the human gaze on the other— but analytic achievement appears only tentative at best.

This expression of epistemologic doubt becomes more apparent if we compare the geologic meditation embedded in the midground of Seymour's image (fig. 58) with the frontispiece to Robert Bakewell's highly influential *Introduction to Geology* (1813), a work we know that James, and so possibly Seymour, consulted during the expedition (fig. 59).[58] This illustration of the summit of Cader Idris in North Wales, which depicts basaltic columns geologically identical to those in *View Parallel to the Base,* is described in the body of the text as "an interesting sketch of a group of columns taken by Henry Strutt, Esq. of Derby, with a *camera lucida,* which may be relied upon as a

57. The image appears in one of Titian Peale's expedition sketchbooks, five of which were donated to the Yale University Art Gallery in 1991 by Ramsay Macmullen, a collateral descendant of the artist.

58. The expedition carried a library of almost one hundred volumes, selected in consultation with Long's scientific colleagues and purchased from Philadelphia bookseller Anthony Finley in two lots in 1819 and 1820. For an annotated inventory and discussion of these works, see Haltman, "Figures in a Western Landscape," 66–68; appendix A, 251–62.

correct representation."[59] The figure of a man in repose at the center of Strutt's image, whether amateur geologist or picturesque tourist, organizes and anchors its jumble of sharp diagonals, suggesting the control exerted by human reason. Bakewell explained this introduction of a human figure into a geologic sketch as providing "a standard of comparison"—the classic application of staffage in geologic

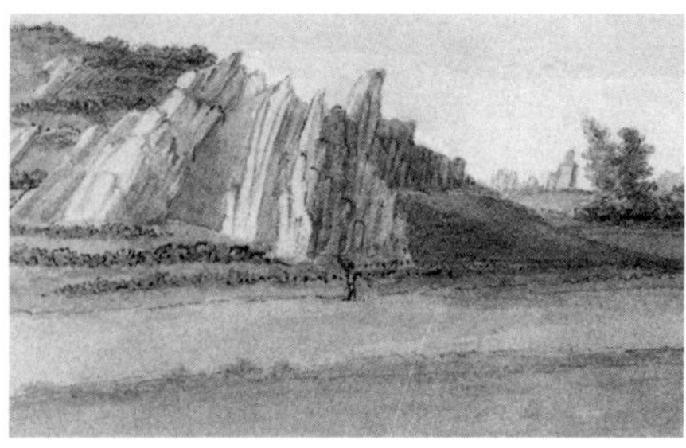

Figure 58. Detail of figure 56, *View Parallel to the Base.*

landscape, as a mensural device. But the figure furthermore represents and so serves to naturalize a specifically *human* standard of comparison and scale. The ideological (here an epistemological) subtext suggests the assertion of rational control over chaotic nature.[60] Seymour's compositional arrangement tends rather to foreground the uncertainty of mensuration, understanding, and control.

In light of James's concerns, this image—along with others Seymour produced that season—serves to portray the field investigations that it represents with remarkable accuracy, even sensitivity. In one sense Seymour's achievement was quite simple: merely picturing the geologic work that he observed—and, more specifically, his colleague's struggle to arrive at geologic understanding— as he saw it, picturesquely. But beyond this, the effortless match between Seymour's compositional technique and geology as subject matter points to the existence of broader structural and even philosophical affinities between these two realms of cultural production. *View Parallel to the Base of the Rocky Mountains* as history painting, as historical portraiture, reveals much about the particular conditions of a single, albeit crucial, field investigation. As illustration it reports the scientific evidence that this investigation gathered; but, as geologic landscape, the image combines these empirical ambitions and transcends them, rendering its occasion—the travails of James, the geologic features of the western landscape—in some sense paradigmatic, thematizing the process of geologic inquiry itself.

59. Bakewell, *An Introduction to Geology*, 297–98.
60. As Solkin has pointed out in his discussion of Richard Wilson's *Cader Idris, Llyn-y-Cau* (ca. 1767–67), such positivist compositions sacrifice sublime potential in the interest of rendering a scene's underlying order and clarity, hence comprehensibility (*Landscape of Reaction*, 224, cat. no. 116).

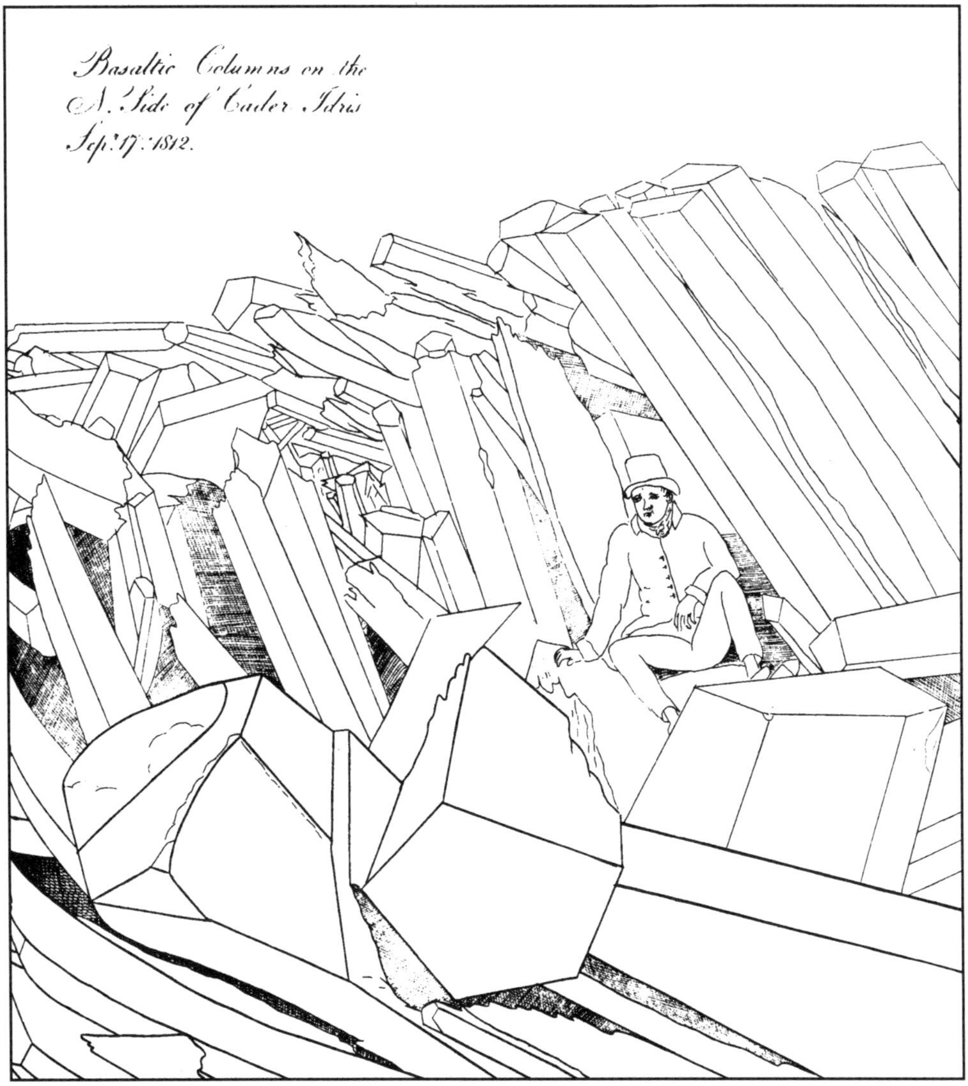

Figure 59. Henry Strutt, *Basaltic Columns on the North Side of Cader Idris* (1813); woodprint, 20.2 x 17. 6 cm. Robert Bakewell, *An Introduction to Geology, Illustrative of the General Structure of the Earth,* 2d ed. (London, 1815), plate 5. University of Southern California.

Despite expectations that new evidence about the earth's origin would emerge from the expedition's close observation of particular formations, neither the Huttonian nor Wernerian theory received decisive support. As an anonymous reviewer had written with regard to the two schools of understanding in Philadelphia's *Analect Magazine* in the spring of 1817, "neither will suffice alone."[61] Arguments in favor of each theory relied heavily on guesswork, and indeed the debate was waged as a skirmish in a wider methodological conflict over the role of hard evidence versus speculation.[62] The discourse of geological science was marked by the same ambivalent devotion to fieldwork—and acknowledgment of necessary recourse to "the wilds of fancy"—captured in Seymour's Long Expedition imagery.[63]

In a review of William Maclure's generally acclaimed *Observations on the Geology of the United States* (1817), Constantine Samuel Rafinesque condemned imaginative speculation outright, declaring that such theorizing constituted "the *novels of geology* rather than its history."[64] In a separate review of Eaton's *Index to the Geology of the Northern States,* Rafinesque went on to propose that "when the speculations of geogony are deduced from history, records, data, remains, analogies, and phenomena, they become a sort of geologic history; but all those which emanate from suppositions, conjectures, fictions, presumptions, probabilities and plausible causes, are at best but ingenious dreams, particularly when they attempt to embrace the origin and the end of our globe."[65] Maclure, who was president of the influential Philadelphia Academy of the Natural Sciences, agreed that the pitfalls for those who would decipher evidence of the earth's origin lay not in speculation per se but in imagination deployed prematurely or in lieu of fieldwork. His *Observations on the Geology of the United States* expressed a pragmatic compromise between the two positions.[66] It was best to begin with the evidence available and then to limit one's conjectures to those stemming from rational analogy: "In all speculation on the origin, or agents that have produced the changes on this globe, it is probable that we ought to keep

61. *Analect Magazine* 9 (April 1817), in a review of Cleaveland's *Mineralogy and Geology,* 312–13.
62. See Martin J. S. Rudwick, *The Great Devonian Controversy: The Shaping of Scientific Knowledge among Gentlemen Specialists* (Chicago, 1985), esp. chap. 2, on the relationship of theory to fieldwork in British geology of the period.
63. In reconstructing the history of this commentary, I have found Robert M. Hazen and Margaret Hindle Hazen's *American Geological Literature, 1669–1850* (Stroudsburg, Pa., 1980) invaluable.
64. *American Monthly Magazine and Critical Review* 3 (1818): 42.
65. Ibid., 177.
66. Maclure, who had studied geologic theory in Edinburgh, famously based his 1817 monograph on fieldwork that took him across the Alleghenies on foot more than fifty times.

within the boundaries of the probable effects resulting from the regular operations of the great laws of nature which our experience and observation has brought within the sphere of our knowledge."[67] Although this qualified endorsement of speculation was classically Wernerian, Maclure carefully distanced himself from Neptunian claims regarding the origins of individual formations. His concerns, as much methodological as theoretical in nature, led him to resolve this debate for himself in eminently moderate fashion: embrace physical evidence insofar as possible, "until we come to the last crust, beyond which we cannot penetrate; then we must drop the thread of positive analogy, and . . . be content with probable conjecture."[68] But some found such conjecture heretical, anticipating the religiously motivated polemics that would become more common later in the century. In 1824, Hayden convened his "Inquiry into Some of the Geological Phenomena To Be Found in Various Parts of America, and Elsewhere"—otherwise a hardheaded survey of the current literature regarding the geology of the United States—with the following pronouncement: "The structure of the globe which we inhabit, and the infinitely varied features, either moral or physical, which are presented to human view in almost every district upon its surface, afford a subject for contemplation, that far transcends the feeble capacity of man perfectly to comprehend." Three installments and four months later, he proposed rather desperately that, while there was *some* worth at least in *both* Wernerian and Huttonian theory,

> in the discussion of their merits, it may be asserted, without
> fear of contradiction, that he who attempts to explain, on the
> principles of either, the infinitely varied phenomena that are
> presented to view, in the structure of the globe; or he who
> attempts to reconcile the equally numerous and varied anom
> alies, glaring inconsistencies, palpable contradictions, and
> inexplicable facts, alike to those principles, will find himself,
> at last, involved in a labyrinth, so inconceivably intricate, that
> it will be impossible to extricate himself except by plunging
> headlong, as many have already done, into infidelity.[69]

In preference to struggling with "inexplicable facts," one could turn either to contemplation in the higher sense, submission to unknowable divinity; or, in the more secular world of mainstream science, to reasoned speculation. The

67. Maclure, *Observations on the Geology of the United States*, iv.
68. William Maclure, "Essay on the Formation of Rocks, or an Inquiry into the probable Origin of their present Form and Structure," *Journal of the Academy of Natural Sciences of Philadelphia* 1, pt. 2 (1818): 266.
69. Hayden, "An Inquiry into Some of the Geological Phenomena," no. 1, 26–27; no. 5, 197.

entry on geology in Gregory's three-volume *Complete Dictionary of Arts and Sciences*, another work in the expedition's library, offered a bemused note: "It must be allowed that those men who, by the mere efforts of their imagination, have endeavored to form ideas respecting the construction, and the great phenomena, of this globe, have numerous titles to our indulgence."[70]

For James at least, the relationship of speculative theory to field observation was a flexible one in the pragmatic mode so vaunted by de Tocqueville[71] and espoused by both Maclure, whom James most quotes in the *Account*, and by Eaton, with whom he had studied. When James described the Floetz Trap formations as being in "appearance and position . . . such as to lead almost involuntarily those least attached to visionary theories into speculations concerning their origin,"[72] his restraint echoes theirs. As Eaton had explained in an essay published a year prior to the expedition, while theory may be deducible from "a great number of facts," without theory "serving as a basis whereon to collate the facts" there can be no understanding of the world's "hidden designs," though these might not themselves prove easily "susceptible to demonstration."[73] Despite voices at each end of the spectrum, most geologists by the time James reached the mountains had embraced some version of this same productive pragmatism, structuring field study in terms of an operative theory and yet remaining flexible enough to modify this conceptual framework wherever necessary in response to anomalous data. It was a compromise singularly well suited to Seymour's own Romantic pragmatism—a reliance on the imaginative reconstruction of observed reality.[74]

We might describe the methodological and theoretical preoccupations of the age as marked by what Kuhn has termed consensual as opposed to crisis science.[75] The participants in the debate were dedicated to the most comprehensive fieldwork—with an obsessively nationalist and positivist enthusiasm for filling in blanks in the geologic map—and at the same time resigned to

70. See George Gregory, *A Complete Dictionary of Arts and Sciences, Including the Latest Improvement and Discovery and the Present State of Every Branch of Human Knowledge*, 3 vols. (Philadelphia, 1816), vol. 2, s.v., "Geology."

71. See Alexis de Tocqueville, *De la démocratie en Amérique*, 2 vols. (Paris, 1835), passim.

72. "Remarks on the Standstone and Floetz Trap Formations of the Western Part of the Valley of the Mississippi," 209.

73. Eaton, "Conjectures Respecting the Formation of the Earth," *Index to the Geology of the Northern States* (1818), appendix, 42.

74. Pointon has described this "alliance between landscape painting and geology" as a "synthesis of a new 'scientific' attitude to observation with a high degree of intensely subjective feeling" ("Geology and Landscape Painting," 87).

75. See Thomas Kuhn, *The Structure of Scientific Revolutions* (Chicago, 1960), passim.

speculative guesswork when it came to the less tangible business of arriving at a comprehensive geognostic understanding. This complicated balance between facts on the one hand and ideas on the other, on material and on abstract realms of understanding, on the particular and on the general, takes form in Seymour's geologic landscapes compositionally and aesthetically as a negotiated middle ground between the correspondingly opposed imperatives of topographic realism (what we see in the middle distance of his watercolors) and a proto-Romantic picturesque (the way this realistic vision foregrounds, even thematizes, its own artifice). Recognition of this compatibility in ways of seeing sheds new light on underlying commonalities between art and science in the early republican age. That these forms of mutually reinforcing cultural practice should be characterized, each in its own way, by a synthesis of similarly competing cognitive modes, transcriptive and inscriptive respectively, leads the way as well to a better historicized and more nuanced understanding of these watercolors themselves, whose governing thematics—origins, origination—determined by the interests of at least two disciplines, can now be seen to resonate in broad, perhaps unexpected, ways.

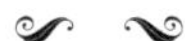

Seymour's *View of Castle Rock on a Branch of the Arkansa[s] at the Base of the Rocky Mountains*, known only in a black-and-white engraving by William Hay in the Philadelphia edition of the *Account* (fig. 60),[76] seems on its surface a conventionally, even formulaic, picturesque response to one of the many unusual geologic formations encountered in the magical terrain through which the expedition found itself passing all that July. James opens his own discussion of this particular *lusus naturæ* using technical language designed to make sense of its strangeness in generic terms: it was another in a class of high perpendicular hills, "sometimes disposed in parallel but interrupted ranges, and sometimes irregularly scattered without any appearance of order." He describes Castle Rock and other similar formations as the "remains" of an underlying bed of sandstone "preserved from disintegration while the contiguous parts

---

76. Owing to editorial difficulties and budget constraints, the American edition of the *Account* contained only ten plates (plus several maps and elevations), eight after works by Seymour. Only two can be considered geologic landscapes: *View of the Castle Rock, on a Branch of the Arkansa[s], at the Base of the Rocky Mountains* (plate 4]) and *View of the Insulated Table Lands at the Base of the Rocky Mountains* (plate 8); I discuss only the former image here.

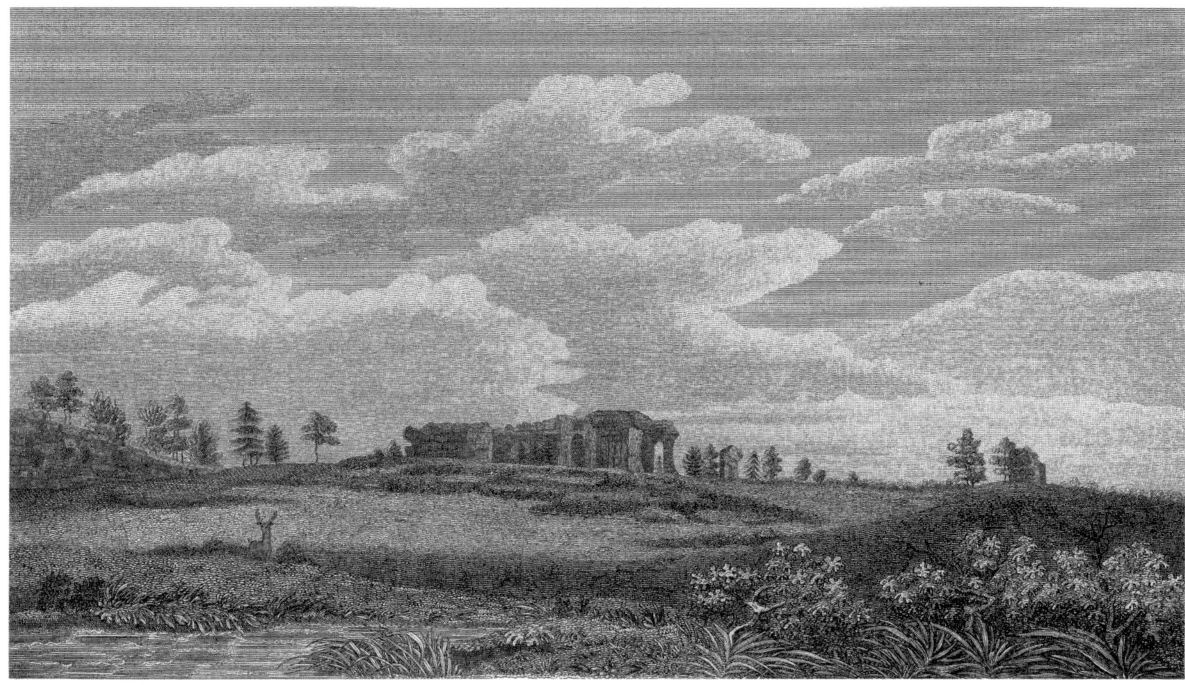

Figure 60. William Hay, after Seymour, *View of Castle Rock on a Branch of the Arkansa[s], at the Base of the Rocky Mountains* (1822); black-and-white engraving, 10.3 x 21.1 cm. Edwin James, *Account of an Expedition from Pittsburgh to the Rocky Mountains* (Philadelphia, 1822–23), plate 4. Huntington Library.

had crumbled down and been washed away."[77] But he goes on in language of a very different sort: "One of these singular hills," James continued, "of which Mr. Seymour has preserved a sketch, was called the Castle rock, on account of its striking resemblance to a work of art. It has columns, and porticoes, and arches, and, when seen from a distance, has an astonishingly regular and artificial appearance." In his diary, James further described this formation as resembling "the ruins of colossal buildings" that "in a country of ruins would certainly be mistaken for a real antique."[78] He was apparently referring not to

77. *Account*, 2:16; 15:306. On the fascination with such geological freaks of nature, see Barbara Maria Stafford, "Rude Sublime: The Taste for Nature's Colossi during the Late Eighteenth and Early Nineteenth Centuries," *Gazette des Beaux-Arts* 87 (April 1976): 113–26; and Stafford, *Voyage into Substance*, passim; see also Rebecca Bedell, "The Anatomy of Nature: Geology and American Landscape Painting, 1825–1875" (Ph.D. diss., Yale University, 1989).
78. *Account*, 2:16; 15:306. James, *Notes*, 1:89. Here, as elsewhere, James echoes the response of Lewis and Clark to similar formations observed two decades earlier a few hundred miles to the north, recorded in Nicholas Biddle, *History of the Expedition under the command of Captains Lewis and Clarke [sic]*, 2 vols.

the temples of ancient Greece and Rome but to European remains from the medieval period. We learn from another member of the party, John Bell, that it was Long who christened this formation Castle Rock (and the rivulet by which it stood Castle Rock Creek), owing in part to the prominence of its "emplacement."[79]

Representation of castles was indeed a staple of picturesque production. Gilpin declared that "among all the objects of art, the picturesque eye is perhaps most inquisitive after the elegant relics of ancient architecture; the ruined tower, the Gothic arch, the remains of Castles, and abbeys. These are the richest legacies of art. They are consecrated by time; and almost deserve the veneration we pay to the works of nature itself."[80] The taste for palatial ruins reflected in Seymour's *Castle Rock* appears earlier in his oeuvre—for instance in his view of Loch Leven (fig. 61), engraved in 1811 for Pinkerton's *Voyages and Travels*. But a painting (or here an engraving) of a naturally occurring simulacrum of a ruin poses a special problem of interpretation. What we gaze upon in *Castle Rock* is not a cultural form overwhelmed by nature like Loch Leven— one of those "vainglorious relics" whose absence was among the distinguishing features of American as opposed to European landscape[81]—but a natural form (residual basalt) with an arguable resemblance to a cultural form (here a castle) overwhelmed by nature. The theme is not *vanitas*, the triumph of death, so much as *mimesis*, the triumph of artifice. The seductiveness of this subject matter—nature simulating ruins—for the artist has to do with its "naturalization" of the powers of dissemblance. Here, in a dizzying originary *mise en abîme*, a cultural form overwhelmed by natural forces is imitated by a natural formation that is then imitated in a work of art—rock castle, castle rock.

(Philadelphia, 1814), yet another of the works the expedition carried. See, for example, in *History of the Expedition*, the description of bluffs along the Missouri by Meriwether Lewis in his entry for 31 May 1805; or, for a modern transcription, *The Journals of the Lewis and Clark Expedition*, ed. Gary E. Moulton, 11 vols. (Lincoln, Neb., 1983–97), 3:225–26. The choice of Seymour's *Castle Rock* for reproduction in the American edition of the *Account* can be explained by the curiosity that it was certain to arouse, but also by the satisfaction it was equally certain to evoke in an audience eager for knowledge of the far West yet aesthetically conditioned to understand grandeur in formal terms as European, as "antique."

79. "Mr. Seymour," Bell noted in his journal on 10 July, "has a well executed view of this, and other subjects seen today"; see *The Journal of Captain John R. Bell, Official Journalist for the Stephen H. Long Expedition to the Rocky Mountains, 1820*, ed. Harlin M. Fuller and LeRoy R. Hafen (Glendale, Calif., 1957), 159–60 (hereafter *Journal*).

80. Gilpin, *Three Essays*, 46.

81. See Laurence Goldstein, *Ruins and Empire: The Evolution of a Theme in Augustan and Romantic Literature* (Pittsburgh, 1977): "To the American democrat the old civilization is properly emblematized by ruins, for the worship of power, the reign of tyranny, the division into class and caste, all of these produced the vainglorious relics that distinguish the European, as opposed to the American, landscape" (p. 218).

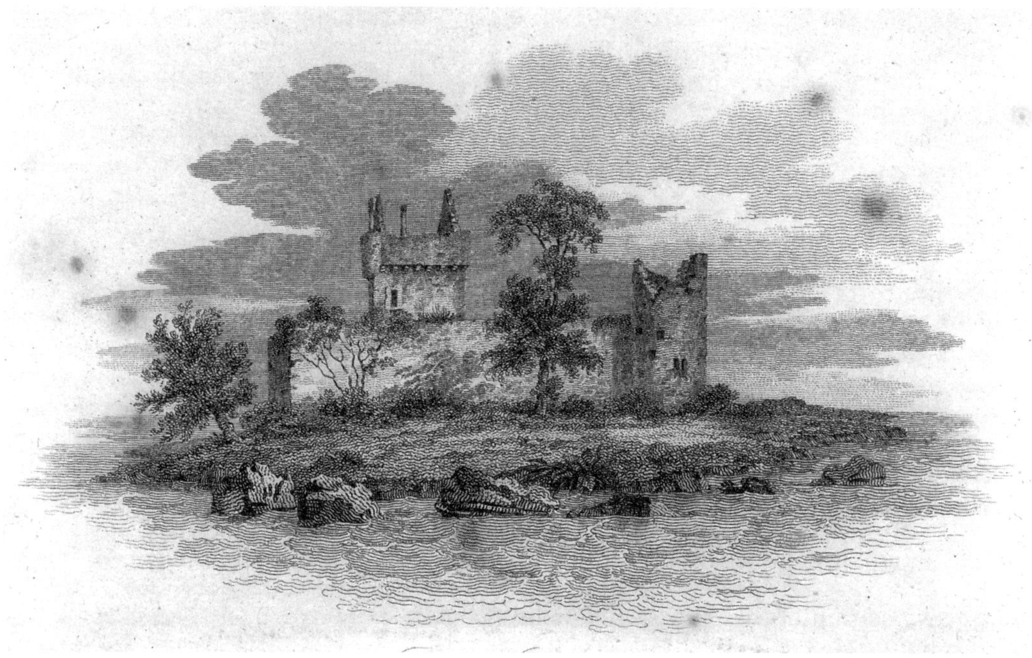

Figure 61. Seymour after George Cooke, *Loch Leven* (1811); black-and-white engraving, 9.2 x 14.5 cm. John Pinkerton, *A General Collection of the Best and Most Interesting Voyages and Travels* (Philadelphia, 1808–13), volume 3. Huntington Library.

In its expeditionary context, Seymour's *Castle Rock* is a form of history painting, serving as a potent visual reminder of the political cataclysm and antiroyalist revolution out of which the nation had emerged. It is a visual allegory featuring an outmoded European institution in decline and, as such, a nationalist recuperation of its symbolic power.[82] Here Nature replaces History as a token of legitimate—that is to say, original—authority. Not a week before, the "great national festival" had been celebrated by an "examination of the mountains" accompanied by "a pint of maize and small portion of whiskey" for each man.[83] In Seymour's hands, the picturesque, itself a kind

82. Seymour's most successful works to date were celebrations of American military victory over the British. His *Brilliant Naval Victory* (1812), a copperplate engraving depicting the victory of the U.S. *Constitution* over the English frigate *Guerriere* on 20 August 1812, and his depiction of Andrew Jackson's victory over British forces in the Battle of New Orleans in 1815, distributed in the form of an aquatint by James W. Steel, both enjoyed wide popularity.

83. *Account,* 1:496; 15:275. Bell's *Journal* contains an exuberant, patriotic account of this commemoration of "the birthday of liberty" (p. 159). For a discussion of early 4th of July celebrations as ritualized republican responses to perceived aristocratic leanings on the part of Federalists, see Simon P. Newman, *Parades and the Politics of the Street: Festive Culture in the Early American Republic* (Philadelphia, 1997).

of English relic, takes on a radical meaning in which old symbols are recast: a "castle" in ruins is made to speak the language of new possibilities.[84] The quest for geologic origins occasioned this depiction; but Seymour's insistence on the originary value of his own art translates this pre-text into an aesthetic signature of his defining presence. The same trope serves the ends of historical allegory as well, figuring the past as origin first toward and then away from which one travels in a quest for meaning.[85]

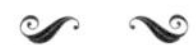

Seymour's *View of the Chasm through which the Platte Issues from the Rocky Mountains* (fig. 62) records the expedition's arrival at the mountains on 6 July 1820. It is best known in its published form as an aquatint by Isaac Clark that appeared as a frontispiece to sometimes one, sometimes another volume of the English edition of the *Account*. We stand before a triangular cliff-face at midground center, formed where two mountains come together; from the base of the cliff, the waters of the South Platte flow in the direction of the picture plane.[86] Although this chasm or narrow gorge is situated at both the thematic and the geomorphic center of the composition, our position with respect to it is left tenuous. The subtle vertical striation of the cliff-face turns to lateral striation on the surface of the water, reinforcing our sense of its flow in our direction, inhibiting movement toward the chasm where the river rises. As in the roughly contemporary *View Near the Base of the Rocky Mountains* (see fig. 55), we are left, seemingly, with nowhere to go. Here, however, our desire to enter picture space is actively resisted by the steady current as well as topographic disruptions and visual discontinuities.

84. On the picturesque as a form of political discourse, see Ann Bermingham, *Landscape and Ideology*, 57–85; see also Bermingham, "System, Order, and Abstraction: The Politics of English Landscape Drawing around 1795," in Mitchell, *Landscape and Power*, 77–101.

85. On the *Map of the Country Drained by the Mississippi* prepared by Long in 1822, based on the expedition's findings and first published in the *Account*, a small drawing of a castle appears at just this point, marking the expedition's farthest westward penetration.

86. James describes the Platte at the foot of the mountains as measuring some "twenty-five yards wide, having an average depth of about three feet; its water clear and cool, and its current rapid" (*Account*, 2:3; 15:289). He mentions elsewhere (1:406; Philadelphia edition only) that a drawing "taken on the spot by Mr. Seymour, of which an engraving has been made for our Journal" displayed "some of the prominent features of this peculiar formation." While an engraving after this work was not in fact included in the Philadelphia atlas, Seymour's drawing did serve as the basis for the aquatint that I discuss here, later published in the London edition.

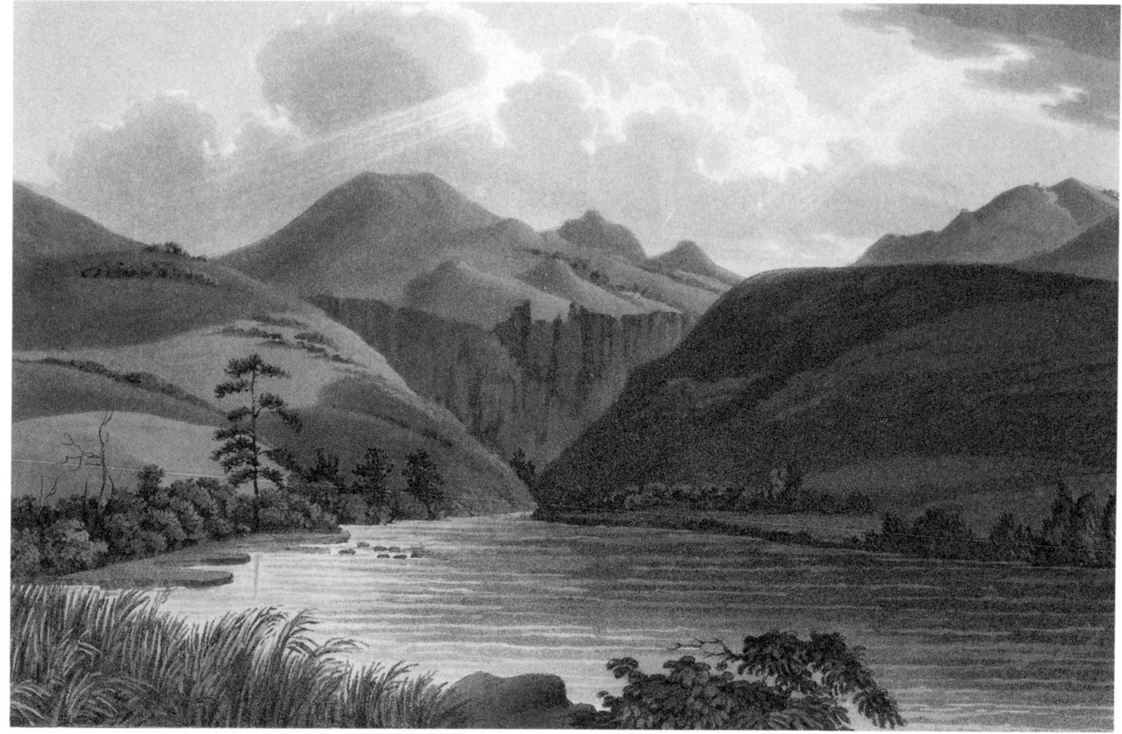

Figure 62. Isaac Clark [after William Hay], after Seymour, *View of the Chasm through which the Platte Issues from the Rocky Mountains* (1823); hand-colored aquatint, 13.3 x 21.4 cm. Edwin James, *Account of an Expedition from Pittsburgh to the Rocky Mountains* (London, 1823), frontispiece to volume 1. Huntington Library.

This visual sense of frustrated entry has its correlative in the uneasy progress of the expedition. "At eleven o'clock," James recounts, "we arrived at the boundary of that vast plain, across which we had held our weary march for a distance of near one thousand miles," there only to discover looming above their heads "a range of naked and almost perpendicular rocks resembling a vast wall."[87] An exploratory party led by James was soon exhausted; he and his men found their way obstructed by "impenetrable and naked rocks," forced eventually to conclude that the barrier was "too hilly and broken to penetrate."[88] His description of the view from below conveys this same sense of thwarted penetration: "This extensive range, rising abruptly from the plain, skirts the base of the mountains like an immense rampart, and to a spectator placed near it, intercepts the view of the still more grand and imposing features of the

87. *Account,* 1:503; 15:285–86; the last line cited is in the London edition only.
88. *Account,* 1:2–7; 15:287–93; and 16:99 (London edition only); see also James, *Notes,* 185.

granitic range beyond."[89] Indeed, this third-person passage conveys the interrupted task of a hypothetical viewer much like ourselves in the presence of Seymour's image, perfectly placed *not* to see what lies beyond the midground. James likens the chasm, beside which the expedition encamped, "to a close line of palisades placed within, and closing the passage which seemed to promise an entrance into the mountains."[90]

While James's frustration resulted from his sense of inability to gain sufficient access to these key basaltic and granitic features, compelling yet elusive evidence of geologic history, Seymour's image, directing one's attention to the sheer cliff-face at midground center, thematizes inaccessibility itself. In this, *View of the Chasm through which the Platte Issues from the Rocky Mountains* strains the conventions of the picturesque, pursuing the strategy of rendering viewer access problematic to such an extent that the aesthetic of controlled irregularity we have seen to govern other of Seymour's expedition images veers dramatically away from the beautiful (with its stress on control) toward the sublime (with its stress on irregularity). Its composition invites in the viewer classically Burkean feelings of frustrated desire, owing to a combination of difficult access, vastness, and the presence of dramatic hidden forces beyond human comprehension. Picturesque landscape painting in this period has been described as embodying an "aesthetic of accommodation" confirming both the artist's and the viewer's sense of being in "interpretive control."[91] At its most basic, the subject matter of such landscape painting is not land itself but the translation of land into landscape that it seems possible to visit and know. Where the conventional structures of control have been impaired, as here, the picturesque becomes the subject more than just the style of representation, and the failure to achieve control itself becomes thematic. The term "chasm" in Seymour's title, taken directly from James, suggests that the headwaters he represents suggest a vortex of barely contained energies.[92]

<hr>

89. *Account*, 2:1, 15:286.

90. *Account*, 2:406 (Philadelphia edition only).

91. Wolf, "Revolution in the Landscape," 214. On the sublime as an "unruly" picturesque, see Martin Price, "The Picturesque Moment," in Frederick W. Hilles and Harold Bloom, eds., *From Sensibility to Romanticism: Essays Presented to Frederick A. Pottle* (New York, 1965), 262–65.

92. In a scientific paper that he presented after the expedition, James offers a more explicit sublime construction of this topos of originary issuance. He cites reports of "a deep and wide chasm in the side of a mountain out of which flames and smoke have often been seen to issue." For this early description of the Yellowstone basin, see Edwin James, "On the Identity of the Supposed Pumice of the Missouri, and a Variety of Amygdaloid found near the Rocky Mountains," *Annals of the Lyceum of Natural History of New York* 1 (1824): 22. Seymour's title is taken from a phrase in the *Account* (1:503; 15:285).

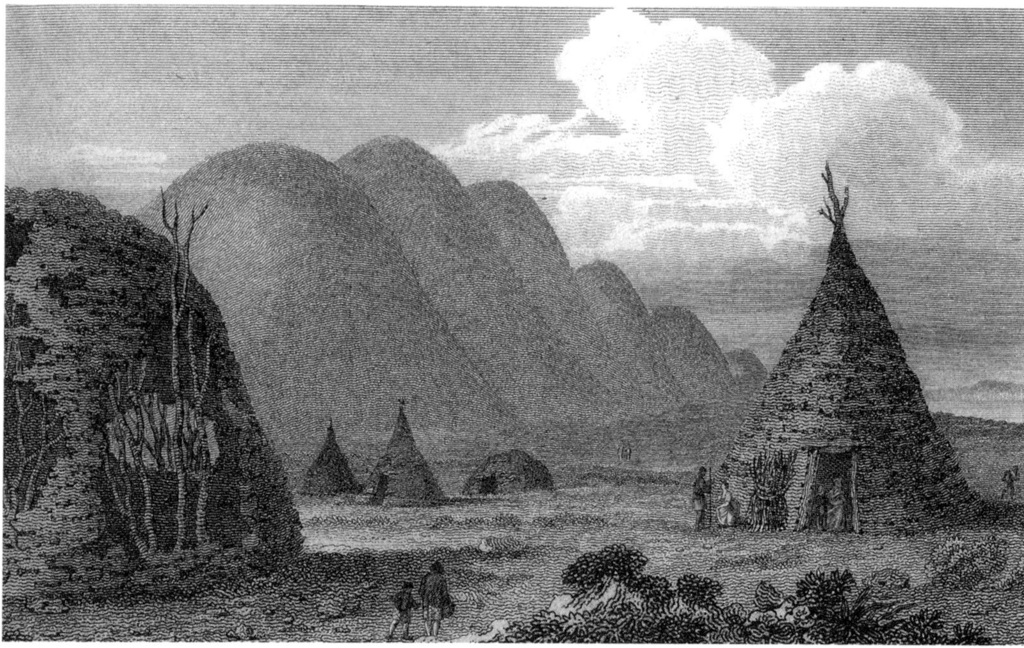

Figure 63. Seymour after George Cooke, *Sheelins in Jura and a Distant View of the Paps* (1811), black-and-white engraving, 8.7 x 14.4 cm. In John Pinkerton, *A General Collection of the Best and Most Interesting Voyages and Travels* (Philadelphia, 1808–1813), volume 3. Huntington Library.

Visual impediments, as elsewhere in Seymour's geologic illustrations, at once compel one to see and yet reduce one to seeing at a distance, making contemplation a necessary virtue. In an essay with the almost humorously cautious title, "Some Speculative Conjectures on the Probable Changes that May Have Taken Place in the Geology of the Continent of North America East of the Stoney Mountains," published shortly after the appearance of the *Account,* Maclure cites the example of "a river running between two precipices of rock in a deep channel" as one that inspired a turn to reverie, an "utmost stretch of imagination," concerning geologic origins.[93] In this he merely echoes James, who had noted: "It is difficult when contemplating the present situation and appearance of these rocks, to prevent the imagination from wandering back to that remote unascertained period, when the billows of a primeval ocean lashed the base of the Rocky Mountains."[94]

93. William Maclure, "Some Speculative Conjectures . . . ," *American Journal of Science* 6 (1823): 98–99.
94. *Account*, 2:2; 15:287. James goes on to speculate about the nature of the great "catastrophe" necessary to explain this cliff-face suspended above a chasm, as a result of which "the secondary ha[d] been broken off and thrown into an inclined or vertical position."

Where James details his failed "opportunities of penetrating and making examinations" within the mountains,[95] his language conflates the concerns of geologist and physician, his other official role during the expedition. Such gynomorphic figuration abounds in the *Account*. Seymour merely pursues the metaphor, finding topographic means to further sexualize the experience of failed penetration and frustrated entry by figuring the chasm as natural vagina, font of originary issuance. The gynomorph is futhermore strikingly complete, with swells of mountain slope at right and left like rounded thighs of earth framing a triangular mons, mountain peaks in the distance uplifted like breasts, and river water breaking like amniotic fluid in a birth canal.[96] In 1820 such usage would have had one likely source in the similarly gendered language of a flourishing travel literature,[97] and another in the plates embellishing those accounts. Seymour had himself engraved at least one such plate, *Sheelins in Jura and a Distant View of the Paps* (fig. 63), published in Pinkerton's *Voyages and Travels* a decade earlier, representing an Alpine mountain range whose name of course already suggested its analogy with female breasts.[98] When

95. James to Eaton, 8 December 1820.
96. The Latin for river is *amnis*. Paul Shepard, in *Man in the Landscape: A Historical View of the Esthetics of Nature* (1967; reprint, 1991, College Station, Tex.), records connections between the female body and landscape (99–100), notably those made by Sigmund Freud, who proposed that, in the language of the dream, the female sexual organs, because of their "complicated" topography, "are often represented by a landscape with rocks, woods, and water," the *mons veneris* a mountain formation located within what he termed the "primary landscape" of the mother's body. Shepard also cites Gertrude Rachel Levy, who notes in *The Gate of Horn; A Study of the Religious Conceptions of the Stone Age and Their Influence upon European Thought* (London, 1948), that a mountain profile is frequently symbolic of more figurative passage, of birth or rebirth. The source of such figuration in aesthetic theory—and the implicitly feminine gendering of "the Beautiful"—is readily apparent from a comparison of Seymour's *View of the Chasm* with Gilpin's *View without Picturesque Adornment* (figure 4, above).
97. Biddle's official history of the Lewis and Clark Expedition, for example (as noted previously, another text carried on the Long Expedition), describes the discovery of the source of the Missouri in the following terms: "They had now reached the hidden sources of that river, which had never yet been seen by civilized man; and as they quenched their thirst at the chaste and icy fountain. . . . they felt themselves rewarded for all their labours and all their difficulties" (Biddle, *History of the Expedition*, 1:359, 12 August 1805).
98. Precedent for such gynomorphism may be seen, for example, in eighteenth-century picturesque landscape gardening in Britain. Sir Francis Dashwood, in perhaps the most notorious example, is said to have "laid out the gardens" of his country house at West Wycombe "by a curious arrangement of streams, bushes and plantation to represent the female form"; see *The Victoria History of the Counties of England*, 5 vols. (1925; reprint, London, 1969), 3:135. William Woollet is said to have captured this curiosity in 1757 in a series of four engravings (including *A View of the Walton Bridge, Venus's Temple, &c. in the Garden of Sr. Francis Dashwood Bart. at West Wycombe in the County of Bucks*) after paintings of the park by William Hannan (see the entry for Francis Dashwood in the *Dictionary of National Biography*); yet his visual allusions prove more elusive than those in *View of the Chasm*—my thanks to Stephanie Ross for these references (personal correspondence, 28 August 1991). Cf. Alexander Pope's "On a Lady" (1713): "But while her pride forbids her Tears to Flow, / The gushing Waters find a Vent below." On the use of anthropomorphic symbols drawn from earlier European allegorical imagery in antebellum painting in the United States, see J. Gray Sweeney, "The Nude of Landscape Painting: Emblematic Personification in the Art of the Hudson River School," *American Art* (fall 1989): 43–65.

Thomas Cole, in 1827, composed a list of possible landscape subjects, he included "a series of pictures called 'Human Life—An Allegory.' The first picture shall be a view of a streamlet issuing from a dark cave & pursuing its course between banks adorned with flowers & graceful trees . . . Over the mouth of the cave will be an inscription in the rock—'Life issues from the womb of dark oblivion.'"[99]

Seymour's geo-metaphoric figure in *View of the Chasm* lends the perception of frustrated penetration an epistemological resonance as well. Our gaze and our imagination are constrained to pass over the surface of the water between river banks, directed to the point of river birthing where all lines converge—precisely where such vision is proscribed. We are thus held at an uneasy distance from the object of desire figured as a woman's body, open-legged and life-giving, an earth mother from which river water issues, Cole's "womb of dark oblivion," a voyeur's sublime. The longing for access to origins so central to the art and the science of the age takes the form here of longing for a "maternal-primitive" that, in Susan Stewart's phrase, "has never existed except as narrative" and thus must ever remain inaccessible."[100]

In historical terms, the potent mix of prohibition and desire at the crux of Seymour's image visually deconstructs the imperial drama in which the expedition played a part. Its implicit narrative discloses the fundamental sense in which the upper Missouri basin as a geopolitical region—economically (as a source of goods, especially furs, and markets) as well as geologically (as a source of information)—originated at the *source* of its westernmost river. The chasm in this respect is the appropriate endpoint to the expedition's "weary march" up the Platte.[101] Knowledge gathered here would address the interests of expansionism symbolically, as a signifier of the west's abundance and, more subtly, as a challenge to those who would possess or merely gauge its riches. Seymour's depiction of this source in *View of the Chasm* plays on the longings for possession that it prompts by sexualizing both the promise and the challenge that the mountains

99. See *Studies in Thomas Cole: An American Romanticist*, ed. Howard S. Merritt (Baltimore, 1967), appendix 2: "Thomas Cole's List 'Subjects for Pictures,'" 88, 92. These works were never executed; but for a related allegory, see Cole's *Childhood*, the first work in his series *The Voyage of Life* (1839–40, oil on canvas, Munson-Williams-Proctor Institute, Utica, N.Y.).

100. Susan Stewart, *On Longing: Narratives of the Miniature, the Gigantic, the Souvenir, the Collection* (Baltimore, 1984), 23.

101. This arrival at the "source" of the Platte comes at the exact center of James's narrative, bridging the last page of the first volume and first page of the second volume of the original edition of the *Account*. This chiastic symmetry is an indication of the self-consciously literary approach taken by James to the expedition's story. On the importance of the Platte as source of information on the geologic origins of the Rockies, see James, "On the Identity of the Supposed *Pumice* of the Missouri," 21–23.

represent; and in his image, the engine of desire is the gaze. James's frustration at the "impenetrability" of the mountains has once again its visual correlative in pictorial (here sublime) irregularity, engaging the eye in the desire to know.

Evelyn Fox Keller has characterized the scientific method itself as the collective means worked out by men of the Enlightenment for dealing with uneasiness and guilt in response to systematic violations of visual (or natural) taboo. Drawing a parallel identical to the one Seymour suggests between the secrets of nature and those of women, Keller concludes that, in modern Europe at least, secrets of both sorts "have traditionally been seen by men as potentially either threatening—or alluring—simply by virtue of the fact that they articulate a boundary that excludes them."[102] The proleptic response to such anxiety has been the presumption by males of the powers of creation, a rescription of "natural processes" that is as typical of scientists as it is of artists. The example of Charles Lyell, the noted British geologist and James's exact contemporary, may be instructive. Rudwick reminds us of Lyell's invocation of the words of historian Barthold Georg Niebuhr: "He who calls what has vanished back again into being, enjoys a bliss like that of creating."[103] Maclure also suggested the creative power of geological theorizing in 1817, in his *Observations on the Geology of the United States*, a book both James and Seymour knew. Maclure proposed that "the pleasure of indulging the imagination" was "superiour to that derived from the labour and drudgery of observation," and this he attributed to "the self-love of mankind . . . flattered by the intoxicating idea of acting a part in the creation."[104]

Seymour too, as we have seen, took pleasure in asserting the "originality" of his artistic vision, which explains in part his interest in this topos of creative issuance. His resistance to the mimetic function of topographic realism was part and parcel of the picturesque conventions within which and against which he worked. But Seymour, in this refiguring of source as chasm, resists the power of artistic precedent as much as he does "truth to nature." The vaginal font through which energy issues forth into the space of representation appears not as a source but as a wall of blockage—not as the culmination of a process

102. Evelyn Fox Keller, "From Secrets of Life to Secrets of Death," in Mary Jacobus et al., eds., *Body Politics: Women, Literature, and the Discourse of Science* (New York, 1990), 178.

103. See Charles Lyell, *Principles of Geology: Being an Attempt to Explain the Former Changes of the Earth's Surface, By Reference to Causes Now in Operation*, 3 vols. (London, 1830–33), 1:74; cited in Martin J. S. Rudwick, "Transposed Concepts from the Human Sciences in the Work of Charles Lyell," in *Images of the Earth*, 68. Rudwick reports that the source of this "sentiment" was the first volume of Niebuhr's *The History of Rome [Römische Geschichte]* (1811–12), trans. J. C. Hare and C. Thirlwall (Cambridge, 1828).

104. Maclure, *Observations on the Geology of the United States*, 14.

(the detailed workings of which it masks) but as *ex nihilo* emergence. The chasm of the birth canal itself, the gap through which the waters of the Platte are said to issue, remains undiscernible in Seymour's rendering. We see instead a picture plane within the picture plane, a sort of *tabula rasa*, as if creation somehow took place at the surface of the image.[105] Thus the process of artistic making is rescripted here by Seymour as a spontaneous emergence of controlled creative impulse, an act of genius, occluding the actual origins not just of his own representation but of the river as well.

For Seymour, by choosing to depict the chasm at a middle distance, from a slight prominence to the southeast, obscured and so mystified this originary flow quite deliberately. James, on his reconaissance, had managed to get a clear view of the chasm from above, discovering that the Platte, to the west of its "emergence"—*beyond* the chasm as it appears in Seymour's image—was formed by the confluence of two smaller rivers, whereupon it "turn[ed] abruptly to the S.E., bursting through a chasm in a vast mural precipice of naked columnar rock." Peale accompanied this party and returned with a sketch.[106] Seymour, however, turned aside this ready-made sublime, finding means instead to picture his pictorial authority. The waters that originate at center midground of his composition, flowing to the east in our direction, represent the process of artistic mediation: a personal experience of topographic fact emerges at the picture plane, remade as new. Such hydrogeologic imagery is visionary, itself a source of reverie through which the viewer's tendency to "wander a little in the wilds of fancy" is encouraged and indulged.

✂   ❧

The history of early republican ideas is marked by an obsessive interest in original creation.[107] A national prehistory had to be formulated commensurate with the promise of the new, and ethnologists were as much involved in this project

---

105. Bryan Jay Wolf has read Frederic Church's 1855 landscape *The Andes of Ecuador* similarly, as creating "a myth of itself as self-begotten"; see "A Grammar of the Sublime, or Intertextuality Triumphant in Church, Turner, and Cole," *New Literary History* 16 (winter 1985): 322.

106. *Account*, 2:6; 15:292.

107. This theme recurs in Washington Irving's *The Sketch Book of Geoffrey Crayon, Gent.* (1819–20), serialized during the expedition, and in James Fenimore Cooper's *The Prairie*, which appeared in 1823, shortly after the party's return, importantly influenced by the *Account*. See Larzer Ziff, *Literary Democracy: The Declaration of Cultural Independence in America* (New York, 1981); and Charlotte M. Porter, *The Eagle's Nest: Natural History and American Ideas, 1812–1842* (Tuscaloosa, Ala., 1986).

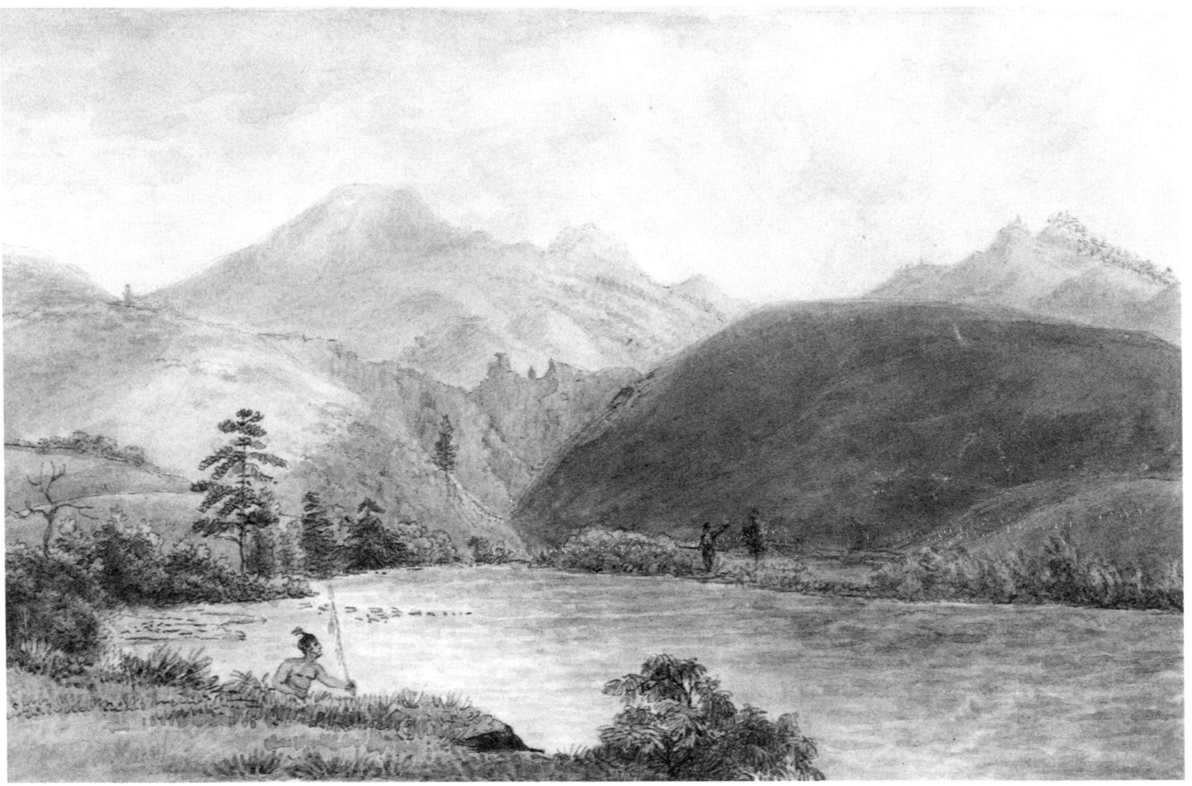

Figure 64. *Sketch on the Upper Mississippi,* or *View of the Chasm through which the Platte Issues from the Rocky Mountains, with Figures* (1822); watercolor over graphite on paper, 12.9 x 20.6 cm. Academy of Natural Sciences, Philadephia.

as geologists. These two concerns were brought together by Seymour in a watercolor known, improbably, as *Sketch on the Upper Missouri* (fig. 64), with the same setting as *View of the Chasm* (see fig. 62).[108] Here, contact between culture and nature—known and unknown—is suggested allegorically in an understated encounter between stylized Indian, to our left, and member of the expeditionary party (perhaps Thomas Say, the expedition's zoologist) to our right, in the looming presence of this same basic gynomorphic configuration.

108. James reports that Seymour executed 150 field sketches over the course of the expedition, 60 of which he "completed" in watercolor in 1822 (*Account,* 2:330n; 17:93n). Of these only 14 are extant today, suggesting (given the importance of its iconographic differences) that the aquatint of *View of the Chasm* (fig. 62), which appeared in the English edition of the *Account,* may have been based on a related watercolor dispatched to the English publisher, Longman, Hurst, Rees, Orme, and Brown, along with other of Seymour's works, and perhaps never repatriated.

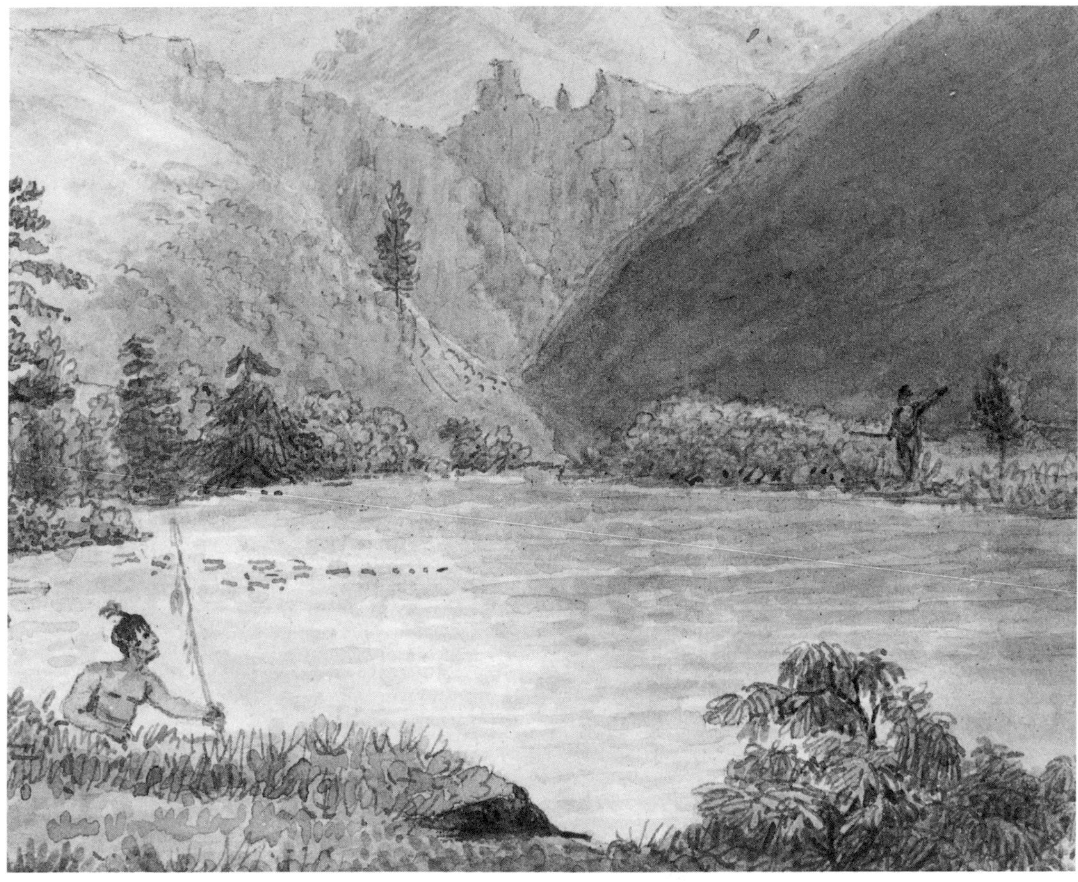

Figure 65. Detail of figure 64, *View of the Chasm, with Figures.*

As is clear from the record, no such event actually took place. Seymour employs this overdetermined figure of the source as the symbolic setting for a myth that he propounds of fresh beginnings in a new world—an Edenic moment of reconciliation across gulfs of difference. By depicting the gap through which the Platte issues from the Rocky Mountains not as "waters bursting through a chasm" but as a smooth uninterrupted flow, emblematic once again of birth and of rebirth, Seymour visually transforms the river's rush between the rocks into a single point of origin, and so refigures what might have been an image of irreconcilable difference—two men on opposite sides of a chasmic divide—into the shallow, continuous curve of what resembles the far shore of a body of still water (fig. 65)—a visual metaphor of possible connection, not unlike "the landscape of a dream."

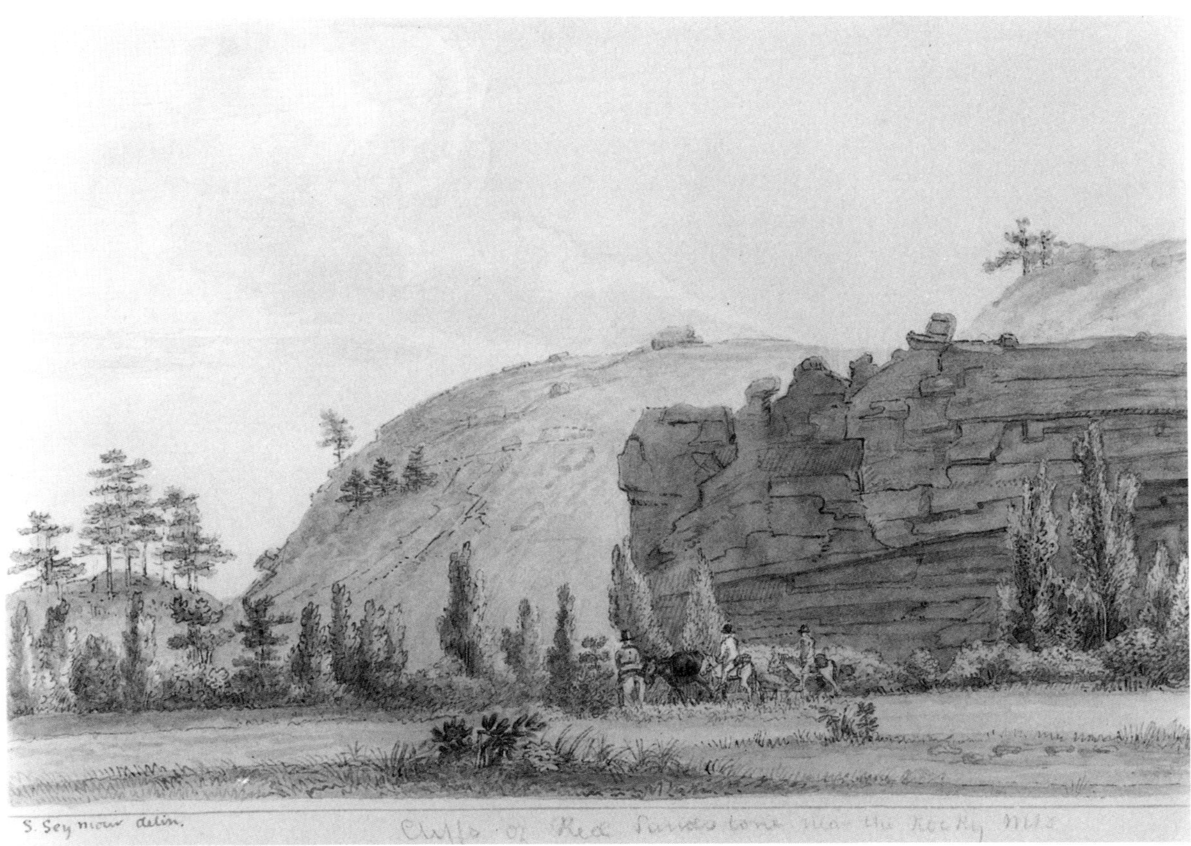

Figure 66. *Cliffs of Red Sandstone Near the Rocky Mountains* (1822); watercolor over graphite on paper. 14.5 x 20.5 cm. Beinecke Rare Book and Manuscript Library.

A similarly dreamlike atmosphere, suggestive of the oneirism underlying geological perceptions, figures in Seymour's *Cliffs of Red Sandstone Near the Rocky Mountains* (fig. 66). This image portrays three men engaged in geological investigation at a considerable distance from the picture plane, accompanied by pack horses or mules. Despite their diminutive size, as in other of Seymour's images these men are not lost within the world he represents but instead provide its focus, located as they are precisely where the composition's three defining spatial axes—the vertical left edge of a large rectangular cliff-face, the horizontal limit of the midground at its base, and the viewer's line of vision—intersect. In Cartesian terms, their presence marks the watercolor's *origin*, at once its central point and source of visual and logical coherence. This underlying structure of coordinate planes is made visible by the presence of these human figures, whose analytic systems and ambitions it reflects.

The presence of scientist-explorers at the composition's spatial origin works once again to reconcile geology and painting, suggesting the family relation between two visionary efforts to make culture out of nature. Seymour's scientists, viewed literally on the move at center stage, engage in this work explicitly, pursuing "truth" into a realm of wilder nature lying off and out of sight at midground left—into the Rocky Mountains, understood as pure unknown—with only reason (and a few hypotheses) to guide them, a symbolic passage that epitomizes Seymour's art as well. Both formally and coloristically, these carefully positioned figures are shown moving from a realm of rectilinear red characterized by observed geological detail to one of green, organic, mammillary smoothness—from a world that has, in some respect, been processed by the mind to one as yet unknown. This movement of science into nature parallels the movement "into nature" of artistic process, what Alexander Pope termed "nature methodized." The eponymous "cliff of red sandstone" gives visual form to the conceptual apparatus governing expeditionary science, with its basis in mathematics, as well as to that governing Seymour's own artistry. Here he employs geometry, or "earth measure," as trope to help us make sense of what we see, the red cliff echoing the watercolor's own rectangularity. The signs of prehistoric cataclysm in the rock face have been quietly reduced to a rectangular blank slate, like the central triangle in *View of the Chasm*, ready for scientific *and* artistic inscription.

Seymour's geologic landscapes would have appealed to readers trained by picturesque convention to appreciate the previously unexperienced as comprehensible when framed and distanced. It is likely that these images would have appealed as well to a populace of amateur geologists prepared to recognize the formations pictured both as keys to the important geogenic questions of the day and as assertions of the centrality of American investigations to world-class intellectual debate.[109] In Seymour's work this question of national identity is posed as a matter of epistemology, the authenticating evidence of original creation being inscribed upon the surface of the world, a nationalist mythos and an aesthetic one as well. As Alan Trachtenberg has noted, "Geological knowledge permits, indeed insists upon, the superimposition of one picture upon

---

109. For a discussion of the function of illustrations in early-nineteenth-century geological reports, see Wagner, "The Idea of Geology in American Landscape Painting," 173–99. Unfortunately, there was essentially no mention of Seymour's illustrations in any contemporary reviews, although Charles Willson Peale both praised and copied several of them. See Charles Coleman Sellers, "Charles Willson Peale with Patron and Populace: A Supplement to *Portraits and Miniatures by Charles Willson Peale* with a Survey of His Works in Other Genres," *Transactions of the American Philosophical Society*, n.s., 59 (May 1969): S119–21.

another—[that of] the invisible past upon the fluctuating image of the present. . . . This additional factor in perception—let us call it geological imagination—results in a doubled vision."[110] Seymour's expeditionary watercolors are scientific illustrations with a difference, for they serve to *thematize* as well as to *illustrate* the work of geology, framing the positivist pretensions of scientific description in terms of picturesque relativity—as the products of "geological imagination."

Michigan State University

110. Alan Trachtenberg, *Reading American Photographs: Images as History* (New York, 1989), 139, 141.

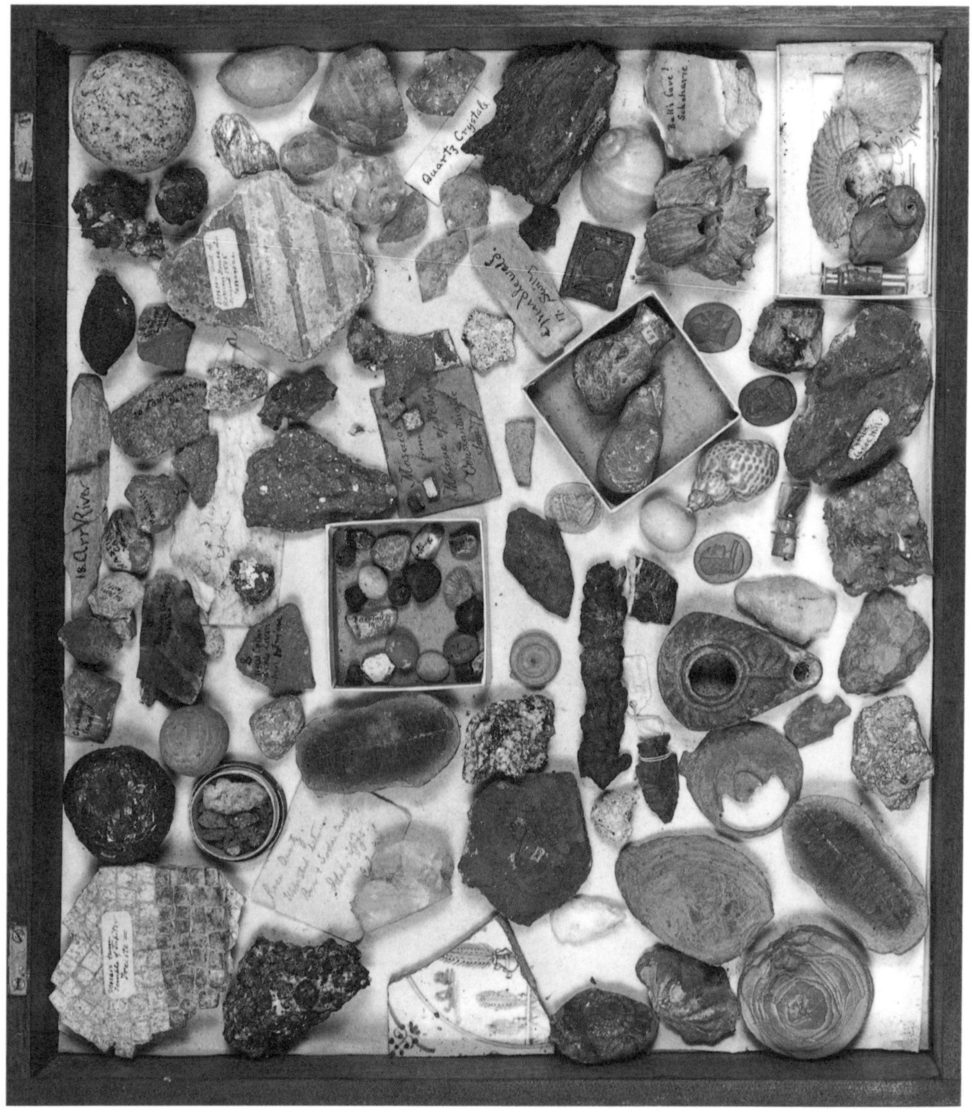

Figure 67. Thomas Cole's fossil and rock collection, 50 x 46 x 7.6 cm. Bronck Museum, Greene County Historical Society, Coxsackie, N.Y., gift of Edith Silberstein Cole.

Thomas Cole and the
Fashionable Science

REBECCA BEDELL

Thomas Cole (1801–48), the leading American landscape painter of the second quarter of the nineteenth century, was a student of geology.[1] His mineral case (fig. 67), now in the collection of the Bronck Museum in Coxsackie, New York, is one of the many bits of evidence we have of his interest in the subject. Measuring only about eighteen by twenty inches, this wooden box contains an intriguing assortment of objects, an assortment reminiscent of the curiosity cabinets of earlier eras. All of the items in the box are of mineral origin, yet some have been shaped and transformed by human hands. Fragments of limestone and quartz reside next to cameos, arrowheads, and a piece of mosaic work.

As his mineral cabinet suggests, Cole made no firm distinction between art and science, between human history and natural history. Just as the makers of the arrowheads and cameos took raw minerals and transformed them into useful objects, so too Thomas Cole sought to do something useful with geological ideas, for instance, teaching lessons about the human condition. The ways that Cole made use of geology in his life and in his art are the subject of this essay.

Cole pursued an interest in geology throughout his career. He began fossil hunting as early as 1822,[2] and he also kept abreast of the latest geological literature. He read Ebenezer Emmons's geological survey of New York State

I wish to thank William H. Truettner, Janet Headley, Susan Danly, Denise Allen, Wendy Greenhouse, Jennifer Danly, and Alex Steinbergh for their thoughtful comments on earlier drafts of this essay. I also wish to express my gratitude to Alan Wallach and Ellwood C. Parry III. These two Cole scholars have not only stimulated my thinking about the artist through their published works but also, over the years, been unfailingly generous in sharing ideas and information with me.

1. On the history of geology in the nineteenth century, see Charles Coulton Gillispie, *Genesis and Geology* (Cambridge, Mass., 1951); and Mott T. Greene, *Geology in the Nineteenth Century* (Ithaca, N.Y., 1982).
2. Cole mentions fossil hunting with his friend William A. Adams in a letter to Adams of May 1838; Cole Papers, New York State Library, Albany (hereafter NYSL), box 1, folder 3 (Archives of American Art [hereafter AAA], roll ALC1).

and John L. Comstock's Geology, a popular work that he kept on his personal bookshelf.[3]

Among Cole's friends and acquaintances were a number of geologists, including Benjamin Silliman, the leading American scientist of his era.[4] Silliman was a professor of geology and chemistry at Yale College; he was also among the first of his countrymen to urge American landscape painters to study geology. In an article written in 1830 he argued that if they did so, the landforms in their paintings would "assume a verisimilitude, depending on physical laws."[5] Cole seems to have followed Silliman's advice, at least to some extent. During his many sketching trips, whether in the northeastern wilderness or through the Italian countryside, Cole filled his journals and sketchbooks with observations on such things as the geological history of Niagara Falls, the effect of erosion on Kaaterskill gorge, and the distinctive characteristics of the lavas of Aetna and Vesuvius.[6] He sometimes annotated drawings made on these trips with geological references, a practice that he followed more often later on in his career. On his *Sketch of Boston from an Old Fortification at Roxbury* of about 1838 (fig. 68), he marked the foreground, "conglomerate of purplish hue." On his *Sketch of Monument Rock near Sand Beach, Mount Desert* of 1844 (fig. 69), he noted "a multitude of granite boulders." Cole's familiarity with geological nomenclature was such that he could use words like *granite* or *conglomerate* as a kind of shorthand. When he was back in his studio, these words could call up the appearance of a particular type of rock, the sparkling surfaces of granite or the rough, pebbly texture of a conglomerate.

3. "Professor Emmons' survey" is cited twice in Cole's papers, once in his 1837 "Catskill, NY" Sketchbook, Detroit Institute of Art (hereafter DIA), (AAA, roll D39, frame 844), and again in a notebook entry for 8 July [1837 or 1838], NYSL, box 6, folder 2 (AAA, ALC3). "Comstock's Geology" is on a list of books in Cole's possession, DIA (AAA, roll D6, frame 265). This book is probably J. L. Comstock's *Outlines of Geology* (Hartford, Conn., 1834). Ellwood C. Parry III and Frank Kelly have argued that Cole was also acquainted with Charles Lyell's famous book *Principles of Geology* (1830–33), and Barbara Novak has pointed out that Cole could hardly have missed the many articles on geology in the periodical literature of the day; see Parry's "Acts of God, Acts of Man: Geological Ideas and the Imaginary Landscapes of Thomas Cole," in Cecil J. Schneer, ed., *Two Hundred Years of Geology in America: Proceedings of the New Hampshire Conference on the History of Geology* (Hanover, N.H., 1979), 60–61; Kelly's "Myth, Allegory, and Science: Thomas Cole's Paintings of Mount Etna," *Arts in Virginia* 23 (1983): 12–13; and Novak's *Nature and Culture* (New York, 1980), 57.
4. On Cole's relationship with Silliman, see Ellwood C. Parry III, "Thomas Cole's Ideas for Mr. Reed's Doors," *American Art Journal* 12 (1980): 37–38.
5. *American Journal of Science and Arts* 18 (July 1830): 210–11, quoted in Ellwood C. Parry III, "Recent Discoveries in the Art of Thomas Cole," *Antiques* 120 (November 1981): 1164.
6. Cole's description of Niagara Falls can be found in his 1828 New York Sketchbook, DIA (AAA, roll D39, frames 193–94). The description of Kaaterskill Falls is quoted in Louis Legrand Noble, *The Life and Works of Thomas Cole* (New York, 1853; reprint, Cambridge, Mass., 1964), 258. On volcanic formations, see "Visit to Volterra," journal entry for 31 August 1831, NYSL, box 4, folder 3 (AAA, roll ALC2).

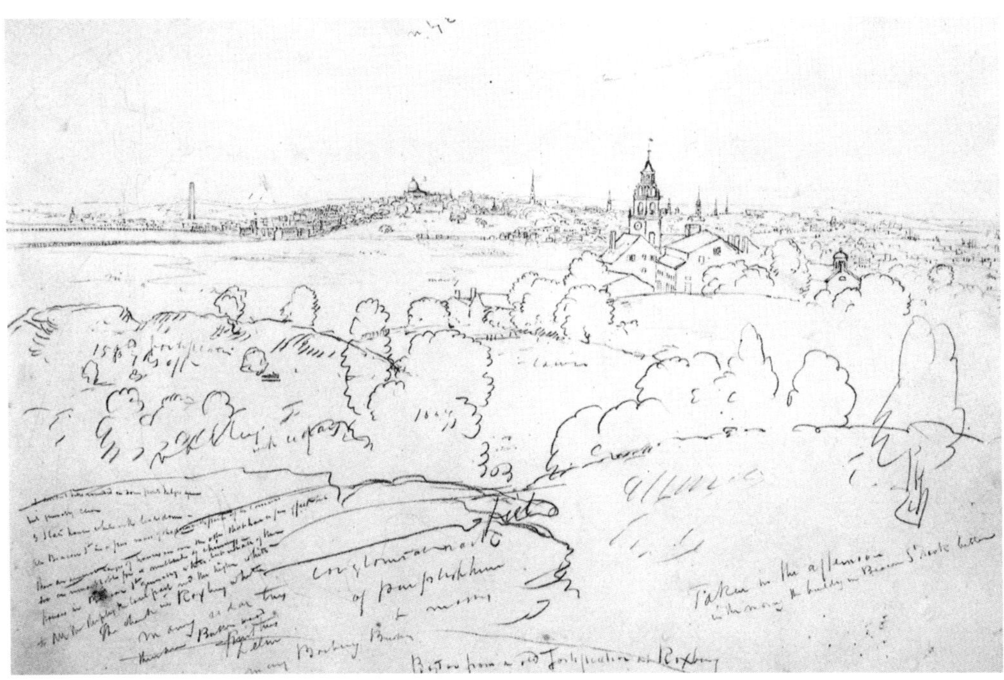

Figure 68. *Sketch of Boston from an Old Fortification at Roxbury* (ca. 1838). Detroit Institute of Art, Founders Society purchase, William H. Murphy Fund.

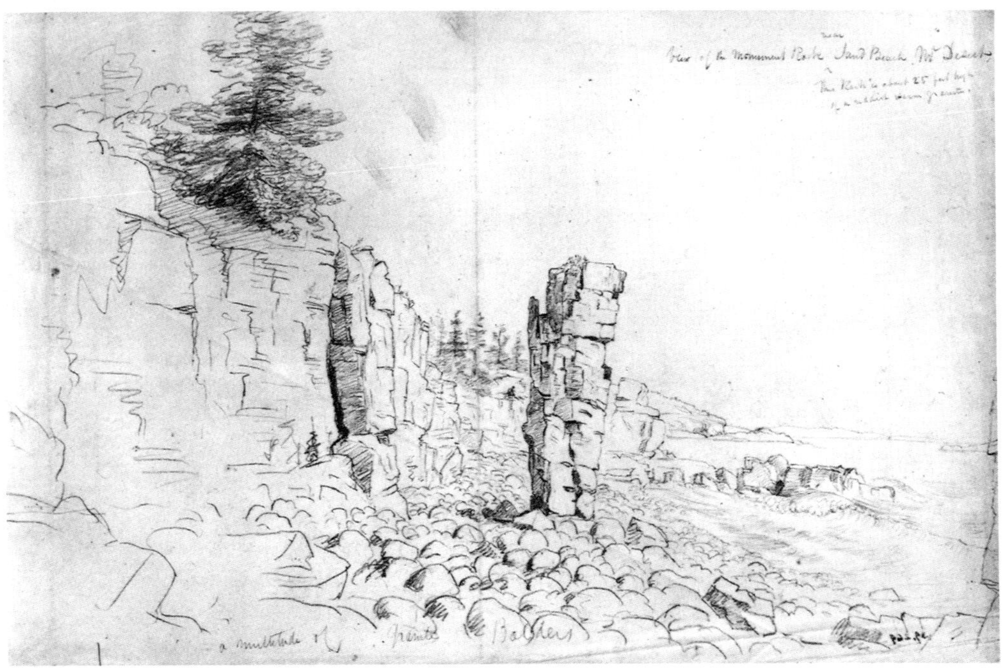

Figure 69. *Sketch of Monument Rock near Sand Beach, Mount Desert* (August 1844); pencil on paper, 24.8 x 40.6 cm. Detroit Institute of Art, Founders Society purchase, William H. Murphy Fund.

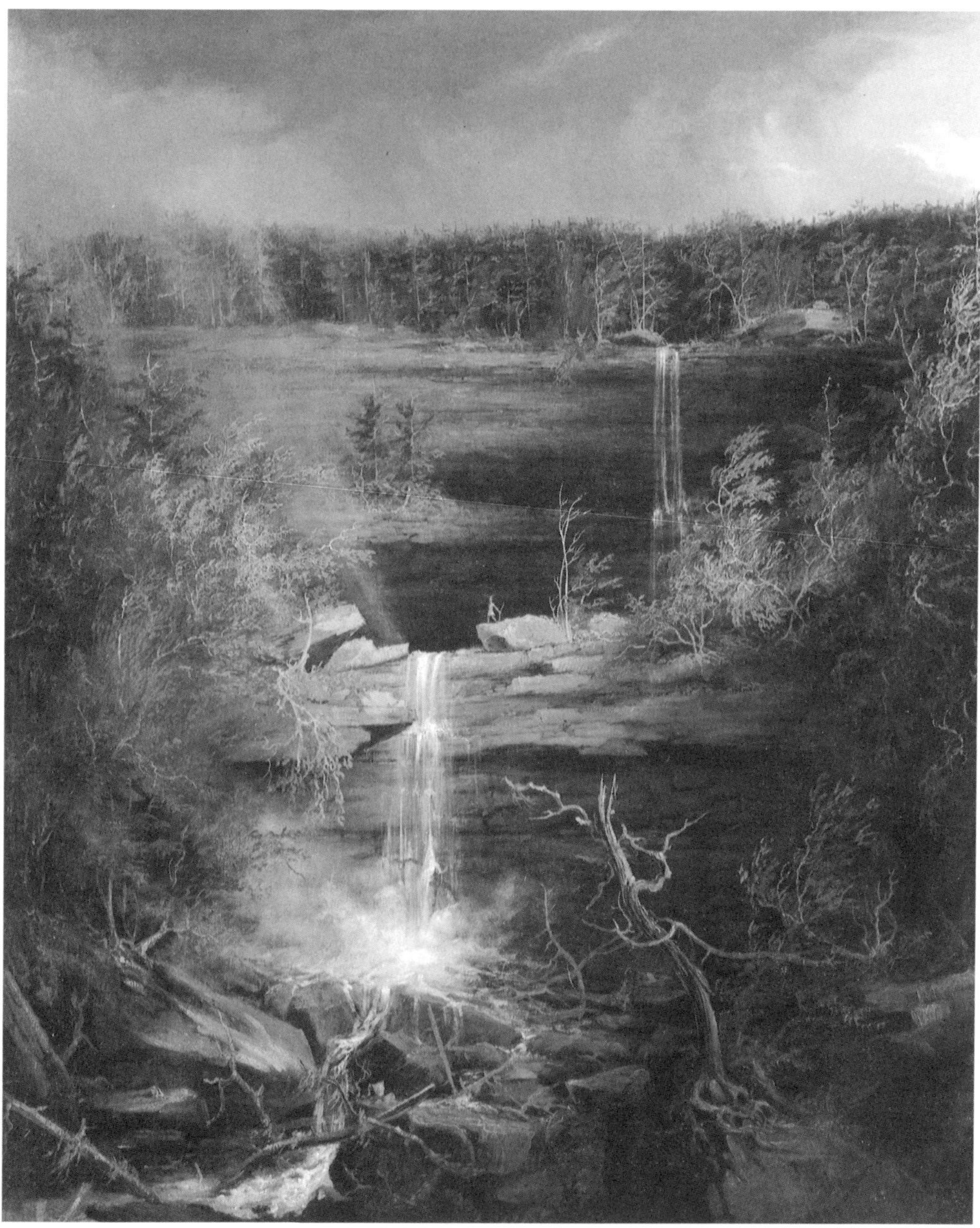

Figure 70. *Kaaterskill Falls* (1826); oil on canvas, 109.2 x 91.4 cm. Warner Collection of Gulf States Paper Corporation, Tuscaloosa, Alabama.

Despite the keen interest in geological facts evidenced in Cole's drawings and journals, his finished paintings are not particularly notable for their geo-

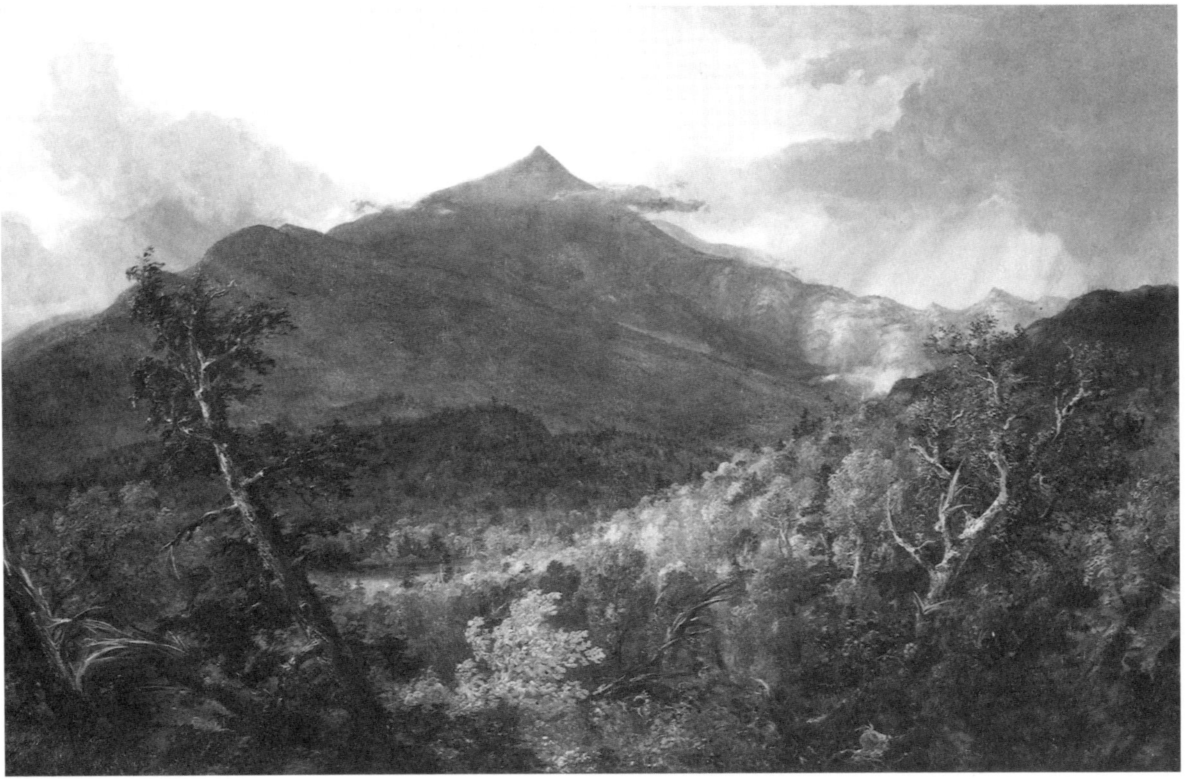

Figure 71. *Schroon Mountain, Adirondacks* (1838); oil on canvas, 100 x 83.8 cm. Cleveland Museum of Art, the Hinman B. Hurlburt Collection.

logic or topographic precision. He often exaggerated the height and distorted the shape of the landforms in his views of American scenery, increasing the height of Kaaterskill Falls, for instance (fig. 70), or, in the case of *Schroon Mountain, Adirondacks* of 1838 (fig. 71), stretching the mountain up into the air, giving it a more strikingly pyramidal silhouette, and sharpening its peak. He also tended to paint his rocks with the same swirling strokes of pink, gray, and tan, whether he was portraying the granites of the White Mountains or the sedimentary formations of the Genesee region.

In his imaginary and allegorical landscapes, as we might expect, he exercised even greater freedom in his handling of geological features, sometimes creating strikingly fanciful formations. In *The Expulsion from the Garden of Eden* of 1828 (fig. 72), for example, he joined a great gothic-shaped archway to a natural bridge in a conjunction that would never be found in nature, while in the *Voyage of Life: Youth* of 1840 (fig. 73) he created a highly improbable cluster of pointed peaks. This nest of needle-like rocks refers not to any

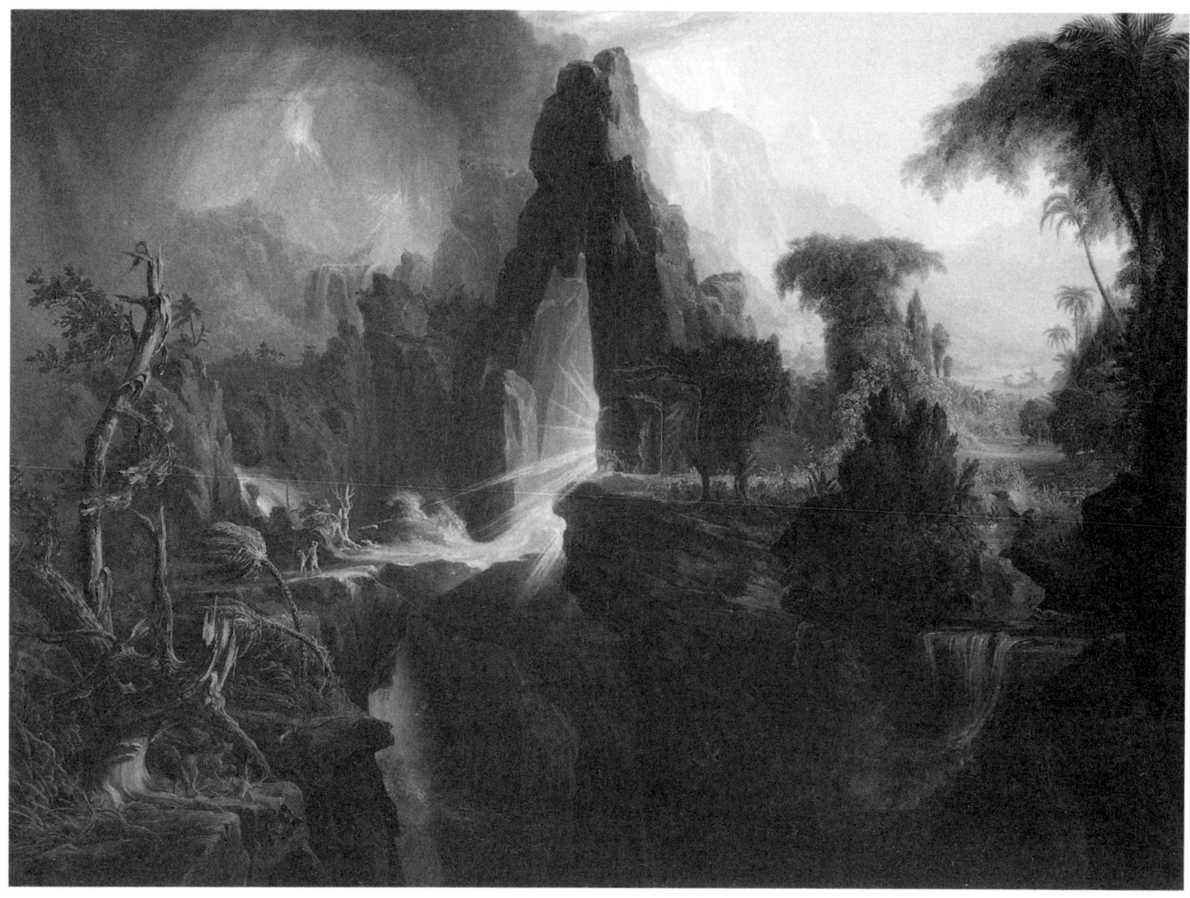

Figure 72. *The Expulsion from the Garden of Eden* (1828); oil on canvas, 99 x 137 cm. Museum of Fine Arts, Boston, M. and M. Karolik Collection.

specific place Cole encountered on his sketching forays but rather to his conception of the fate awaiting human beings as they advance through life. Traveling along the river of life, the young man in the painting is still surrounded by soft and verdant scenery, but the barren, unscaleable peaks in the background suggest the trials he will eventually encounter. As his treatment of these landforms implies, Cole was usually less concerned with fidelity to natural fact than with symbolic and associative potential.

What, then, did geology mean to Cole? Why did he study the science? Certainly by the 1830s and 1840s he had come to regard the study of natural sciences, including geology, as essential *preparation* for landscape painters. It served to familiarize them with the subject of their art, much as the study of human anatomy served portraitists. In February of 1840 he wrote in his

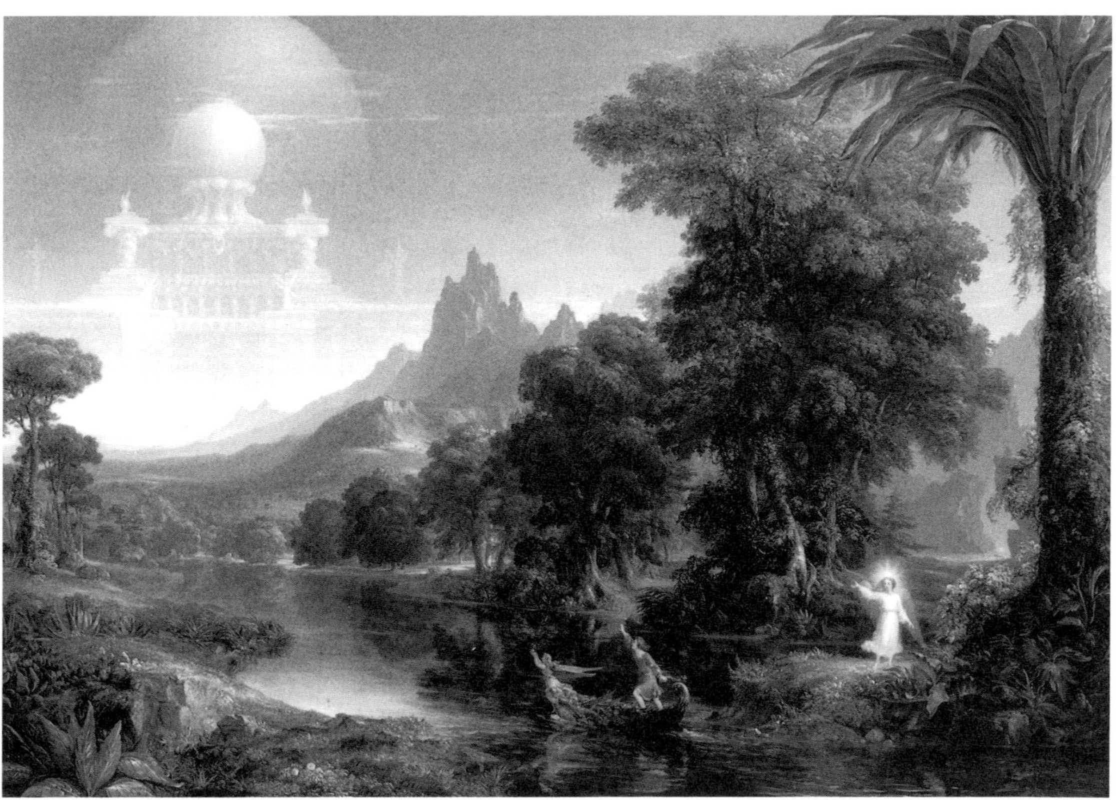

Figure 73. *Voyage of Life: Youth* (1840); oil on canvas, 133 x 200 cm. Munson-Williams-Proctor Institute, Utica, N.Y.

journal, "It is absolutely necessary that the painter have a minute I may say an anatomical knowledge of nature as well as a general one."[7] Yet, deeply influenced by neoclassical art theory—especially the writings of Sir Joshua Reynolds—he also believed that the painter should generalize his representations of the natural world rather than particularize them. The landscape painter, he thought, should study the natural world in all its specificity, but before transferring scenes to canvas he should, as he told his friend Asher Durand, "wait for time to draw a veil over the common details, the unessential parts, which shall leave the great features, whether the beautiful or the sublime, dominant in the mind."[8] For Cole, the specificities of rocks and landforms seem often to have fallen into the category of "common details" and "unessential parts."

7.   From the sketchbook "Thoughts and Occurences," NYSL, box 6, folder 2 (AAA, roll ALC3).
8.   Quoted in Matthew Baigell, *Thomas Cole* (New York, 1981), 13.

But the study of geology, I believe, did more for Cole than simply prepare him for the close observation of nature. Geology served Cole not only artistically but also socially. His study of the science bolstered his claims to gentry status; it was knowledge that separated him from what his patron Philip Hone called "the vulgar and uneducated masses."[9] As Dixon Ryan Fox pointed out long ago, this was a time when the vast majority of Americans could be divided into two groups, gentry and commoners, and it was very important to Cole, raised in England and accustomed to its class system, that he be identified with the former.[10] His study of geology—an intellectual attainment associated with the cultural elite—served him in this regard.

In 1834 a writer for *Knickerbocker* magazine proclaimed that geology "is indeed the fashionable science of the day; and may be said to form a necessary part of practical and ornamental education."[11] Just how fashionable geology was at the time, just how necessary it was to the education of a gentleman, is indicated by the number of Cole's patrons who studied it. In the 1820s and 1830s, the majority of his most important patrons were devoted to the discipline, some as amateurs, some as professionals. Most of these men were members of the old Federalist elite. This group was losing ground politically in the early nineteenth century and turned increasingly to intellectual and cultural attainments— including the study of geology—to justify its elevated social position.[12] In 1807 the incorporators of that elite cultural institution, the Boston Atheneum, admonished their members, "Let men of leisure and opulence patronize the arts and sciences among us; let us love them, as intellectual men."[13]

Cole launched his career as a landscape painter in New York City in 1825. Among the first purchasers of his landscapes was the Federalist Daniel Wadsworth, Connecticut landowner and founder of the Wadsworth Atheneum in Hartford.[14] Wadsworth was Benjamin Silliman's brother-in-law. He was also an amateur geologist and amateur artist whose sketches of geologic features were occasionally reproduced on the pages of the *American Journal of Science*. This journal, the major American scientific periodical of Cole's day, was edited by Silliman.

9.   *Diary of Philip Hone*, ed. Allan Nevins (New York, 1936), 445.
10. Dixon Ryan Fox, *The Decline of the Aristocracy in the Politics of New York* (New York, 1919); on Cole's relationship to the English class system, see Alan Wallach, "Thomas Cole and the Aristocracy," *Arts Magazine* 56 (November 1981): 94–106.
11. Samuel L. Metcalf, "The Interest and Importance of Scientific Geology as a Subject for Study," *Knickerbocker* 3 (April 1834): 227.
12. On the history of this group, see Fox, *Decline of the Aristocracy;* and Frederick Cople Jaher, *The Urban Establishment: Upper Strata in Boston, New York, Charleston, Chicago, and Los Angeles* (Urbana, Ill., 1982).
13. Quoted in Jaher, *Urban Establishment*, 35.
14. Wadsworth's geological interests have been discussed in Parry, "Recent Discoveries," 1162.

Cole's relationship with Wadsworth was a close and affectionate one, and Wadsworth was especially paternalistic toward the young artist in the mid-1820s.[15] Cole visited Wadsworth at his estate Monte Video as well as at his Hartford house, and the two seem to have shared many country walks together. Perhaps on these outings Wadsworth conveyed to Cole something of his enthusiasm for geology. At any rate, Wadsworth was only the first wealthy aristocrat of Cole's acquaintance with an interest in the subject.

The English-born George William Featherstonhaugh (1780–1866)—who, unlike Wadsworth, was a professional geologist—was another of Cole's early patrons. A frequent contributor of papers to scientific societies on both sides of the Atlantic, he also founded a scientific periodical, the *Monthly American Journal of Geology and Natural History.* He intended it to rival Silliman's *American Journal of Science* but it was short-lived. He also authored several geological reports for the United States Government.[16] In November 1825, near the very beginning of Cole's career, Featherstonhaugh made him an offer in the long-established tradition of aristocratic patronage: a painting room for the winter at his estate Featherston Park, near Duanesburg, New York.[17] In exchange for room and board Cole was to paint for Featherstonhaugh a number of pictures at the relatively low price of twenty or thirty dollars apiece. Cole gratefully accepted the offer at the time, but he looked back on this experience as one of the most mortifying of his life. Instead of treating Cole as a social and intellectual equal, Featherstonhaugh seems to have regarded the artist more as hired help, as a bought-and-paid-for symbol of his own aristocratic status. He supposedly shunted Cole off to paint in an unheated room during the day and tried to consign him to dinners with the children at night. This treatment severely bruised Cole's ego. At the same time, it must have impressed upon him some sense of the relative social status of scientists and artists.

Moving back to New York City in early 1826, Cole met more congenial patrons, including the Baltimore merchant Robert Gilmor. Gilmor was an avid amateur geologist, a founding member of the American Geological Society, and, along with Silliman, was elected one of its vice presidents.[18] Over fifty

---

15. On Cole's relationship with Wadsworth, see *The Correspondence of Thomas Cole and Daniel Wadsworth,* ed. J. Bard McNulty (Hartford, Conn., 1983).

16. On Featherstonhaugh's contributions to American geology, see George P. Merrill, *The First One Hundred Years of American Geology* (New Haven, Conn., 1924), 136–39.

17. Information on Cole's relationship with Featherstonhaugh is drawn largely from Ellwood C. Parry III, *The Art of Thomas Cole: Ambition and Imagination* (Newark, N.J., 1988), 28–30.

18. The officers of the American Geological Society are listed in *American Journal of Science and Arts* 2 (1820): 141. See also Chandos Michael Brown, *Benjamin Silliman: A Life in the Young Republic* (Princeton, N.J., 1989), 310. I am indebted to Alan Wallach for calling my attention to the connections between Gilmor, Silliman, and the American Geological Society.

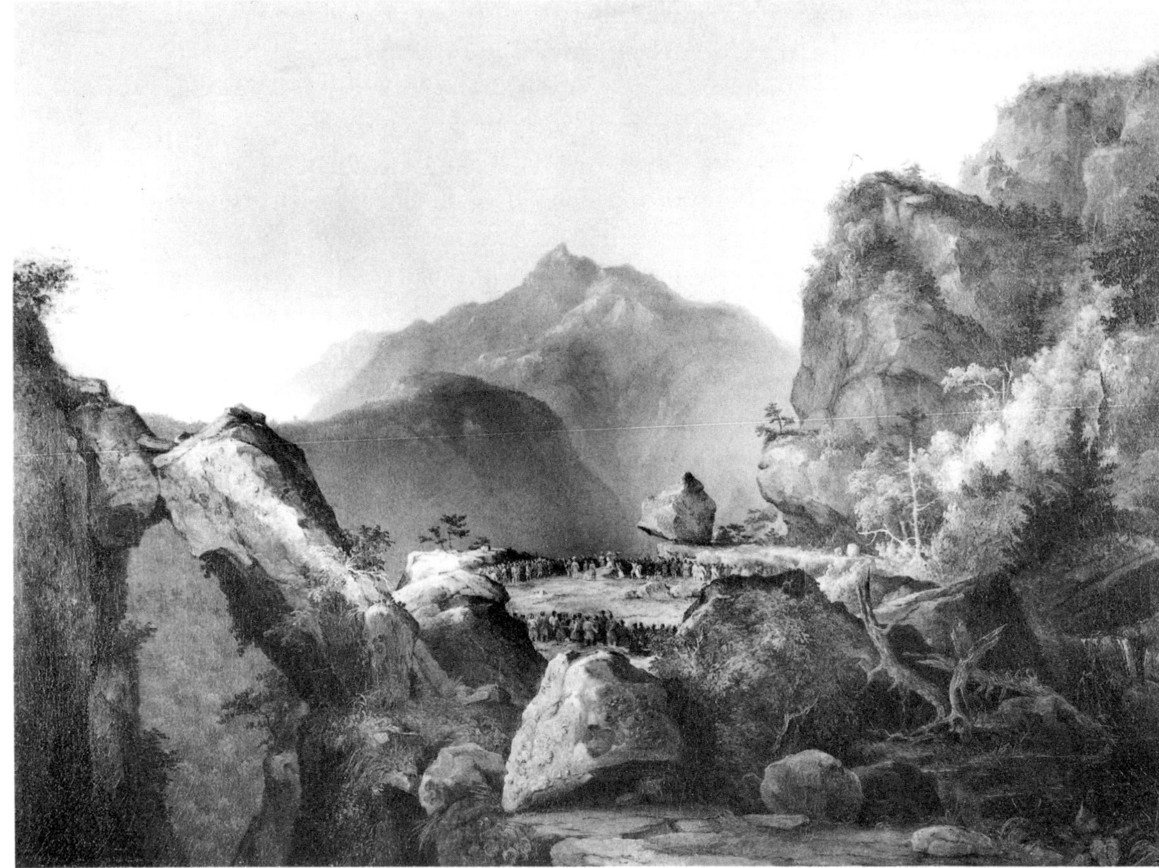

Figure 74. *Scene from "The Last of the Mohicans": Cora Kneeling at the Feet of Tamenund* (1827); oil on canvas, 63.5 x 78.7 cm. New York State Historical Association, Cooperstown, N.Y.

volumes on geology lined the shelves of his library, including *Silliman's Geological Lectures* and *Featherstonhaugh's Geological Report.*[19] One day in December 1827, Gilmor uncrated a box from Cole containing two pictures, *View of Corroway Peak in New Hampshire—Sun Setting* and *The Last of the Mohicans* (fig. 74).[20] Gilmor quickly dashed off a letter to Cole informing him of their arrival and offering his critique of the images. He was, in general, pleased by the pictures, but he pointedly criticized Cole's handling of the geological features in *The Last of the Mohicans.* "The arrangement of the rocks," he told Cole, "is *artificial:* the rock falling against another on the opposite side of a deep ravine, is finely rendered, but strikes the eye as something forced, & not

19. See *Catalogue of the Library of the Late Robert Gilmor* (Baltimore, 1849). Copy in the library of the Peabody Institute, Baltimore. I am indebted to Janet Headley for bringing this catalogue to my attention.
20. Parry, *Art of Thomas Cole,* 62.

ordinarily seen in nature." He was also bothered by the large rock in the center of the composition. "The rock teetering on its pivot," he informed Cole, "is scarcely *balanced* enough. The spectator looks for it falling."[21]

As I will argue later in greater detail, Cole chose these geological formations primarily for their dramatic and thematic potential—aspects that seem to have escaped Gilmor, at least upon his first acquaintance with the painting, when his scientifically trained eye led him to judge the elements of the composition in terms of their fidelity to natural appearances. What I want to suggest here is that criticism like this, indicating that his patrons were closely examining the geological formations in his paintings, must have motivated Cole to pursue his studies more seriously. And despite the overarching concern that I have noted with the symbolic potential of landscape, his handling of geological features became increasingly naturalistic in the years following this contretemps with Gilmor.

Gilmor fancied himself an informed connoisseur who could offer useful advice to a young artist. In Luman Reed (1785–1836), however, Cole found a patron with whom he enjoyed greater rapport, a patron who was at once more generous with his money and less stinting with his praise, a patron who looked up to Cole as an accomplished painter rather than treating him as a struggling young artist in need of help and guidance. Between 1833 and 1836 Reed commissioned ten works from Cole, including the *Course of Empire* series (figs. 78–83).

Like many New York gentlemen of the era, Reed pursued an interest in geology. He owned forty volumes of Silliman's *American Journal of Science* and built a gallery in his New York house to display not only pictures by many of the leading American artists of the era but also cases of minerals.[22] Reed purchased his mineral collection from the Austrian consul general, Baron von Lederer, who was then departing for Europe.[23] It was one of the major private mineral collections in the United States, one well known to practicing geologists.[24]

By most standards of the day Luman Reed was a successful man, but his social position was rather different from that of Wadsworth, Featherstonhaugh, and Gilmor, who were born to wealth and social privilege. Reed was a self-made man. He began his career as a store clerk, but by the age of forty-five he had amassed a fortune in the dry goods business and had turned the daily

21. Quoted in Parry, *Art of Thomas Cole*, 64.
22. See Ella M. Foshay, "Luman Reed: A New York Patron of American Art," *Magazine Antiques* 138 (November 1990): 1074–85.
23. Mrs. Jonathan Sturges, *Reminiscences of a Long Life* (New York, 1894), 158–60, quoted in Parry, *Art of Thomas Cole*, 178; see also Ella M. Foshay, *Mr. Luman Reed's Picture Gallery* (New York, 1990), 39.
24. See Ethel M. McAllister, *Amos Eaton: Scientist and Educator* (Philadelphia, 1941), 276.

operations of his company over to his associates. Reed then devoted himself to the collecting of art and the pursuit of science—his entrées to the fashionable world. In this regard, Baron von Lederer's mineral collection, which combined the prestige of science and the pedigree of European nobility, must have seemed to Reed an especially fortunate acquisition.

Cole, like his patron Reed, was greatly concerned with his social status, and both seem to have pursued the study of geology to distinguish themselves socially. As Alan Wallach has pointed out, Cole was obsessed with class, a preoccupation that had its roots in his childhood.[25] He was born in the English Midlands into the family of a relatively well-to-do textile manufacturer, but when Cole was still a boy his father's business failed. He was then forced to leave school and go to work as an engraver in a textile factory. The horrors of the industrial workplace and the indignities of working next to what his friend and biographer Louis Noble called his "rude fellow operatives" left an indelible impression on him. For the rest of his life he was haunted, as Wallach has said, by a "fear of sliding into the working class."[26] In his adult years, it was essential to his sense of self-worth that he be identified as a gentleman, as a member of the social and cultural elite. With so many of his aristocratic patrons involved in the study of geology, Cole must have recognized early in his career the social utility of geological knowledge.

Cole's concern with rank and status shaped his view not only of the social world but also of artistic hierarchy. Landscape painters, as he lamented over and over again in his journals, letters, and lectures, were not properly appreciated by the critical and academic establishments. Landscape painting was ranked below history painting, below portraiture, and only slightly above cattle pieces.[27] He complained of this slight in a letter to his friend, the painter C. R. Leslie, who wrote to Cole from London in May 1835: "I quite agree with you that Landscape does not rank high enough in the scale of art as established by modern criticism."[28]

Because history painting was the most highly regarded of the genres, Cole was much at pains to elevate his landscapes to that level (and himself to the status of history painter) by introducing into them the sort of epic themes and moralizing lessons that were then associated with history painting. The geology

25. Alan Wallach, "Thomas Cole and the Aristocracy," *Arts Magazine* 56 (November 1981): 94–106. My account of Cole's childhood and of the formation of his class consciousness is based on this article.
26. Quoted in Wallach, "Thomas Cole and the Aristocracy," 95.
27. See, for example, the listing of genres by rank in "The Exhibition of the National Academy of Design, 1827," *United States Review and Literary Gazette* 2 (July 1827): 244; quoted in Parry, *Art of Thomas Cole*, 52–53.
28. C. R. Leslie to Cole, London, 11 May 1835; NYSL, box 2, folder 4 (AAA, roll ALC1).

embraced by Cole and his patrons was ideally suited to this task. At a time when many European geologists, including Cole's nemesis Featherstonhaugh, were insisting on an empirical approach to the science emphasizing the mapping of strata and the classification of minerals, Cole and the majority of his patrons adhered to a more conservative and traditional geology.[29] Theirs was a patriotic and moralistic science bound to revealed religion—a science, in short, designed to convey the sort of moral and spiritual lessons, demonstrated with historical examples, that Cole sought to incorporate into his art. This view of science is epitomized in the person and writings of Benjamin Silliman.

Like many of Cole's patrons, Silliman was a member of the old Federalist elite.[30] The son of a general who had served in the Continental Army during the revolutionary war, Silliman made a socially advantageous marriage to Harriet Trumbull, daughter of Jonathan Trumbull, Federalist governor of Connecticut, and niece of the painter John Trumbull, president of the American Academy of Fine Arts. Silliman's social connections and political affiliations together with his intellectual attainments allowed him to move in the highest of American social circles. Few of Cole's patrons were unacquainted with Silliman, even those with little interest in geology. Take, for example, Philip Hone, who purchased two of Cole's landscapes in the mid-1820s. A one-term mayor of New York City and an ardent Whig who was acquainted with a number of American presidents, Hone pursued a wide range of interests beyond politics, including theater, literature, opera, history, and good food and wine. In the several thousand pages of a diary that Hone kept from 1828 to 1851, he expressed little interest in science, yet when Silliman arrived in New York in 1836 to offer a course of lectures on geology, Hone was in attendance. And when Hone arrived in New Haven for a Whig convention in 1840, Silliman escorted him and a number of his colleagues to Yale College for a quick glance at what Hone described as "the splendid mineralogical cabinet."[31] Cole was probably introduced to Silliman by his early mentor John Trumbull, or perhaps by his patron Daniel Wadsworth (Silliman's brother-in-law). They became close enough that Cole visited Silliman at his home in New Haven and corresponded with him on a number of occasions.[32]

29. On Featherstonhaugh's views, see his *Geological Report of an examination made in 1834 of the elevated country between the Missouri and Red rivers* (Washington, D.C., 1835).
30. Information on Silliman drawn primarily from Brown's biography.
31. *Diary of Philip Hone*, 209.
32. See letters from Thomas Cole to William A. Adams dated March, October, and December 1839 requesting fossil specimens for Silliman (NYSL, box 1, folder 4); and letter from A. N. Skinner to Cole dated September 1839 mentioning a visit by Cole to Silliman (NYSL, box 3, folder 1).

Silliman was a pious Christian and committed patriot who placed his science in the service of God and his country. He founded the *American Journal of Science* in 1818, as he said, to encourage the discovery and exploitation of "those physical resources, which the bounty of God has given us."[33] He also sought to uncover the intersections between science and theology, between Genesis and geology, and this was of particular interest to Cole. Silliman's *Geological Lectures*, published in 1820,[34] dealt with a variety topics, but in particular with the geological evidence of the Flood.

Silliman, along with many of his scientific contemporaries in the United States and Europe, believed that geological evidence of the Flood was readily apparent all around the globe. In the northeastern United States, proponents of this view pointed to great U-shaped valleys like New Hampshire's Carrigain Notch and Crawford Notch, seeing them as thoroughfares excavated by rushing waters on a vast scale. They identified great piles of unsorted rocks, sand, and gravel (such as those that make up Cape Cod) as deposits left by the receding waters and called them "diluvial drift." Of the various categories of diluvial drift, "erratic boulders"—rocks that have been deposited some distance from their place of origin—received particular attention at the time. Some of them can be found in prominent positions on hillsides and mountaintops; these were referred to in Cole's day as "perched boulders." Other erratics were left stranded atop other rocks, some balanced so delicately that they can be rocked by the human hand; these were dubbed "rocking stones." All of these phenomena— U-shaped valleys, diluvial drift, and erratic boulders—are now attributed to the glacial action of past ice ages, but it was their associations with biblical history that captured people's imaginations in the mid-nineteenth century.

Cole was among those who became fascinated by these diluvial features, and he incorporated them (particularly erratic boulders) into a number of his paintings of the 1820s and '30s.[35] They appear in his allegorical-literary works, including *The Last of the Mohicans* (figs. 74, 77) and the *Course of Empire* series (figs. 78–83), and also in his views of American scenery such as *The Notch in the White Mountains* (1839; fig. 84). Cole could certainly have heard about the proposed connection between these geological features and the Deluge through friends and patrons such as Silliman, Gilmor, and Wadsworth. He could have read about them in the *American Journal of Science*, which carried

33. Quoted in John C. Greene, "Protestantism, Science, and American Enterprise: Benjamin Silliman's Moral Universe," in Leonard G. Wilson, ed., *Benjamin Silliman and His Circle* (New York, 1979), 17–18.
34. *Outline of the Course of Geological Lectures, Given in Yale College* (New Haven, Conn., 1820).
35. Parry, in "Acts of God, Acts of Man," notes Cole's inclusion of erratic boulders in *The Last of the Mohicans* paintings and the *Course of Empire* series. He does not, however, draw a connection between these stones and the Deluge theory.

dozens of articles, some illustrated, about erratic boulders, rocking stones, and diluvial drift.[36] By 1834 Cole could also have read of the geological proofs of the Deluge in his personal copy of Comstock's *Geology*, a book that takes as its central theme the reconciliation of geology with revealed religion. In a long section devoted to the Flood, Comstock places great emphasis on diluvial boulders and drift.

> The effects of that grand and awful cataclysm are still to be traced in every country, and in nearly every section of . . . the globe. Vast accumulations of rounded or waterworn pebbles, huge blocks of granite, and immense beds of sand and gravel are found in places where no causes now in operation ever could have have placed them; and still that they have been moved is evident from the circumstances or places where they occur.[37]

A number of Cole's sketchbooks from the 1820s and 1830s contain drawings of what appear to be rocking stones, perched boulders, and other unusually situated rocks. For example, in one sketch from 1831 a flat-topped, oblong boulder teeters atop a rocky ledge, and in another, a similarly shaped rock is precariously balanced on two rounded stones (fig. 75). That same year Cole made a similar drawing that he labeled "Rock lying on the side of a mountain." This slight sketch of a boulder on a slope depicts the sort of geological phenomenon that was widely cited as evidence of the Flood.

Cole was almost certainly familiar with the biblical interpretation of these geological phenomena; and he was without question fascinated by the Deluge itself. In 1825 he wrote a poem called "The Deluge: A Judgment" in which he described the fate of "the last sad victim of the rising flood."[38] A few years after penning this morbid verse, he made a sketch in his notebook that he inscribed "the Deluge." It depicts several figures standing atop a conical mound of earth as torrents of water pour from the heavens and huge waves threaten to engulf them. In the same notebook, Cole also listed potential subjects for future works, including "First Rainbow after the Deluge" and "World after the Deluge."[39] Sandwiched between these are two other subjects that may have been intended as depictions of postdiluvial phenomena: "Rocks and trees heaped confusedly together as having been carried by the floods from the

36.   Parry, *Art of Thomas Cole*, 64.
37.   J. L. Comstock, *Outlines of Geology* (Hartford, Conn., 1834), 80.
38.   NYSL, box 5, folder 2 (AAA, roll ALC3).
39.   Sketchbook inscribed "New York 1827," NYSL, box 6, folder 3 (AAA, roll ALC4). Although the sketchbook is dated 1827, it covers several subsequent years.

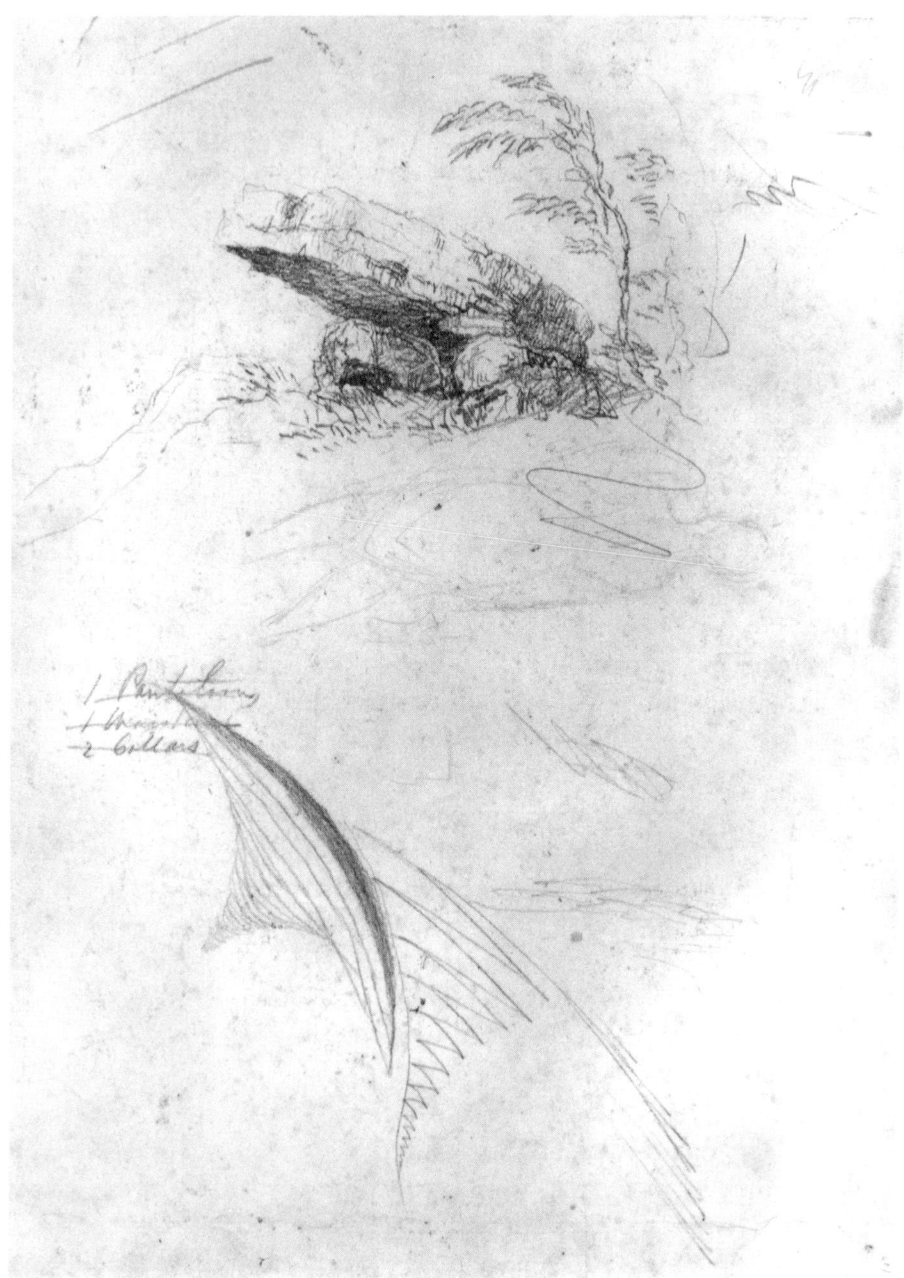

Figure 75. *Drawing of Rocks* (1831). Detroit Institute of Art, Founders Society Purchase, William H. Murphy Fund.

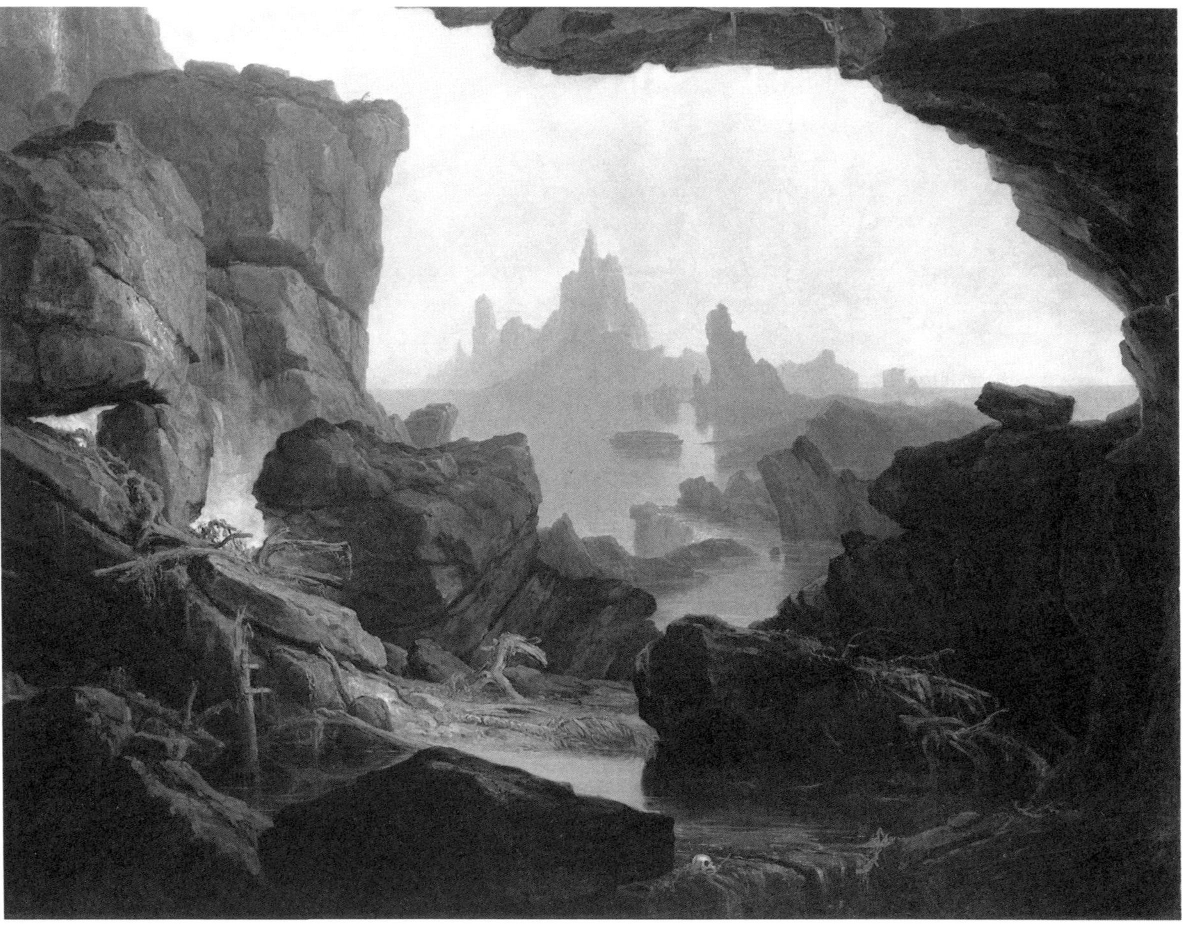

Figure 76. *The Subsiding of the Waters of the Deluge* (1829); oil on canvas, 91.4 x 122 cm. National Museum of American Art, Smithsonian Institution, Washington, D.C., gift of Mrs. Katie Dean in memory of Minnibel S. and James Wallace Dean; museum purchase through the Smithsonian Collections Acquisitions Program.

mountains" and "A picture in which nothing shall be but bare rocks and clouds—Rocks piled on rocks."

In 1829, at about the same time that he made this list, Cole produced a major painting of the Flood itself, *The Subsiding of the Waters of the Deluge* (fig. 76). From beneath a rocky overhang, the viewer looks out on a watery yet peaceful world. The rains have finally ceased, and the golden light of dawn suffuses the scene; the ark is floating serenely in the distance, and the dove is skimming over the waters. The human skull and shattered shipmast in the foreground and the chaotic assemblage of rocks, however, recall the nightmarish destruction of the previous days. The jumbled masses of rocks

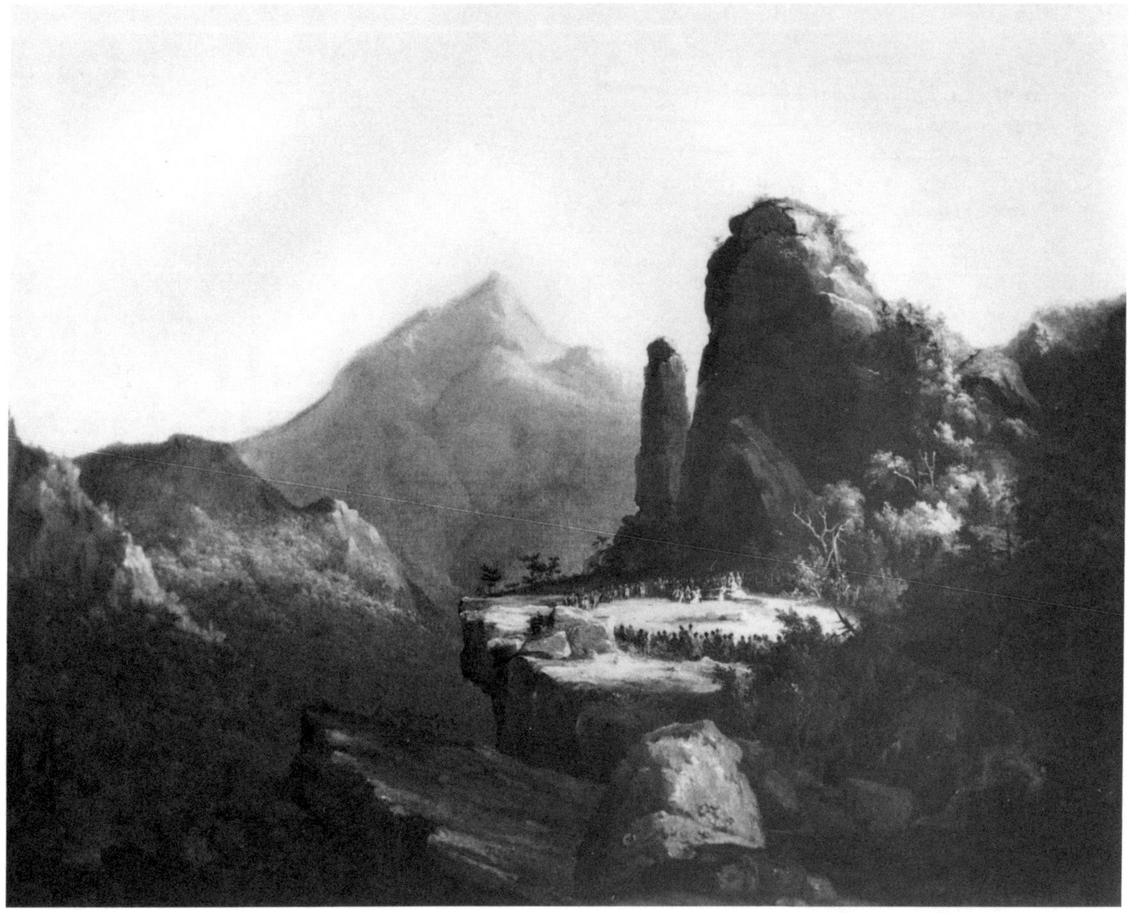

Figure 77. *Scene from "The Last of the Mohicans": Cora Kneeling at the Feet of Tamenund* (1827); oil on canvas, 64.4 x 89 cm. Wadsworth Atheneum, Hartford, Conn., bequest of Alfred Smith.

in the foreground and boulders balanced on mountain peaks in the distance are geological features typically associated with the Flood. This painting was purchased by Dr. David Hosack (1769–1835), a fashionable New York physician and an accomplished botanist, yet another of Cole's wealthy patrons with an interest in geology.[40]

For Cole, the Flood was a historical occurrence of profound moral and religious significance, and features such as erratic boulders were tangible proof

40. Hosack owned one of the earliest mineral collections in the United States, and he also collected fossils that he forwarded to the Royal Academy of Science in London. See Stephen Dow Beckham, "Colonel George Gibbs," in Wilson, *Silliman and His Circle*, 31; and "Letters of Mr. Brongniart with Remarks," *American Journal of Science* 3 (1821): 225. In the late eighteenth and early nineteenth centuries, fossils were believed by many to be remnants of the Flood; see Gillispie, *Genesis and Geology*.

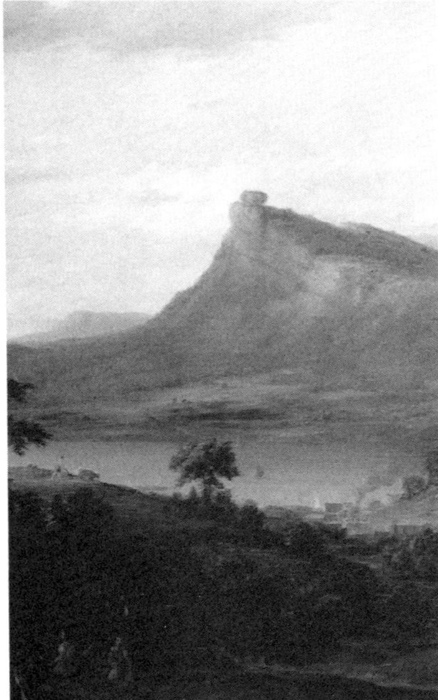

Figure 78. Detail of the boulder from *The Pastoral State* (1834), figure 80. Collection of the New-York Historical Society.

that it had actually taken place. They attested the credibility of the Bible, affirming its truths, and functioned as reminders of the power and potential wrath of God. Erratic boulders made their first appearance in Cole's art in his several versions of *The Last of the Mohicans*, and comparing two versions painted in 1827 helps to show how Cole used the stones as both a moral and compositional feature. In the first of the two, painted for Daniel Wadsworth, he set a rounded erratic atop a stone pillar (fig. 77). In a version painted later that same year for Robert Gilmor, Cole perched a large angular boulder (perhaps a rocking stone) on a ledge in the middleground of the painting. While the boulder in the earlier version could be overlooked fairly easily, the one in the second is much larger and more prominent compositionally. Its striking presence in the center of the canvas calls attention to the tiny figures beneath, which might otherwise be overwhelmed by the landscape elements. The precarious, threatening position of the boulder, overhanging and overshadowing the central figures, also alludes to the nature of the unfolding human drama. Cora, the young white woman, is begging for mercy from Tamenund, the chief of the Delawares. In the ensuing action she will be killed, along with Uncas, the last of the Mohicans. By juxtaposing the diluvial boulder with this violent scene, Cole was perhaps intending to draw a parallel between the deaths of Cora and Uncas (who represented the last of his race) and the destruction of human life in the Deluge. In other paintings by Cole, boulders are similarly linked with the demise of a social unit: a race, a tribe, an empire, a family.

In the *Course of Empire* series, commissioned by Luman Reed and painted between 1833 and 1836,[41] an erratic boulder (fig. 78) appears perched on a

41. Much has been written about *The Course of Empire*. Among the most important treatments of the series are Angela Miller, "Thomas Cole and Jacksonian America: *The Course of Empire* as Political Allegory," *Prospects* 14 (1989): 65–92; Alan Wallach, "Thomas Cole: Landscape and *The Course of Empire*," in Truettner and Wallach, *Thomas Cole*, 23–112; Wallach, "Cole, Byron, and *The Course of Empire*," *Art Bulletin* 50 (December 1968): 376; and Parry, *Art of Thomas Cole*, 137–87.

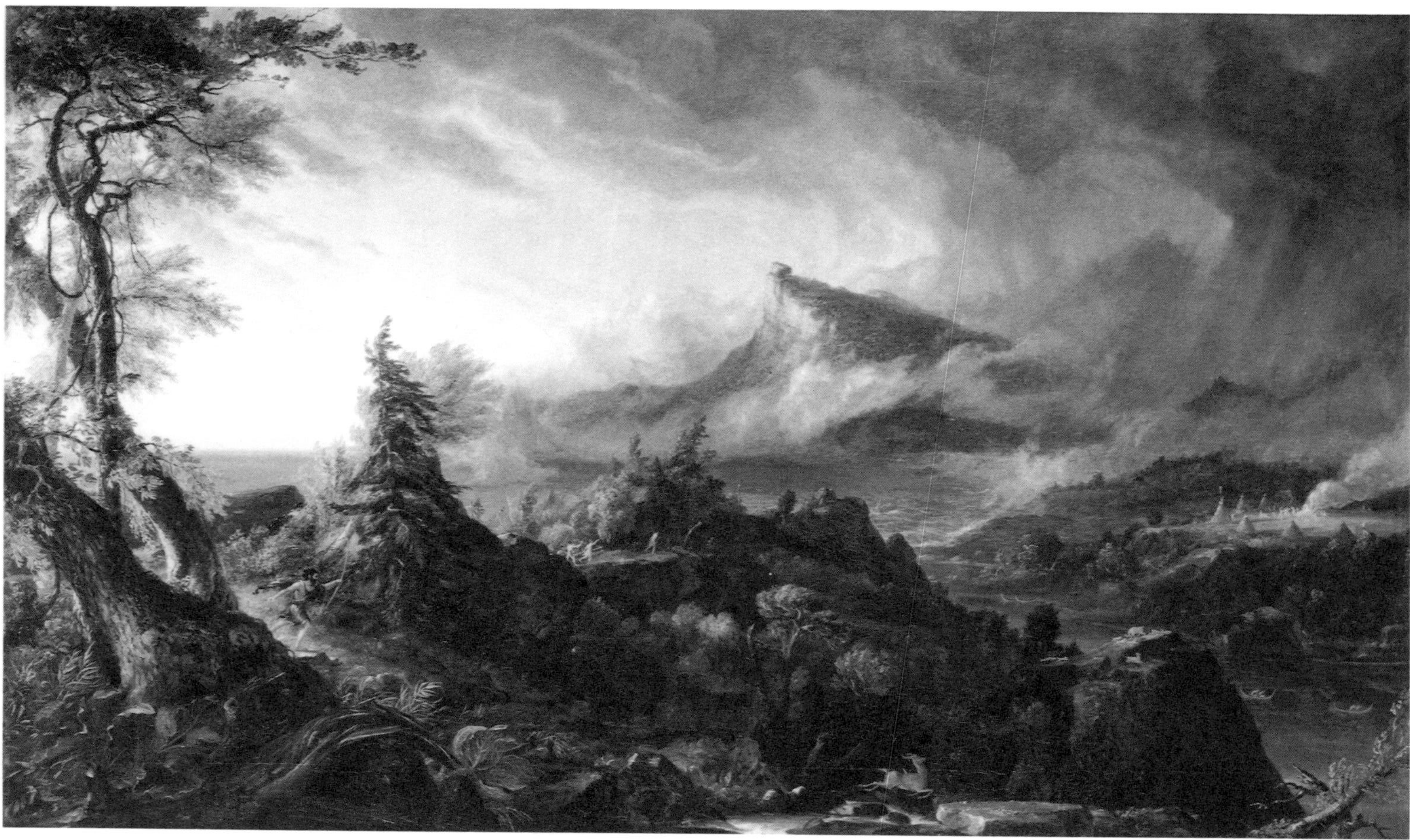

Figure 79. *The Course of Empire: The Savage State* (1834); oil on canvas, 100 x 161 cm. Collection of the New-York Historical Society.

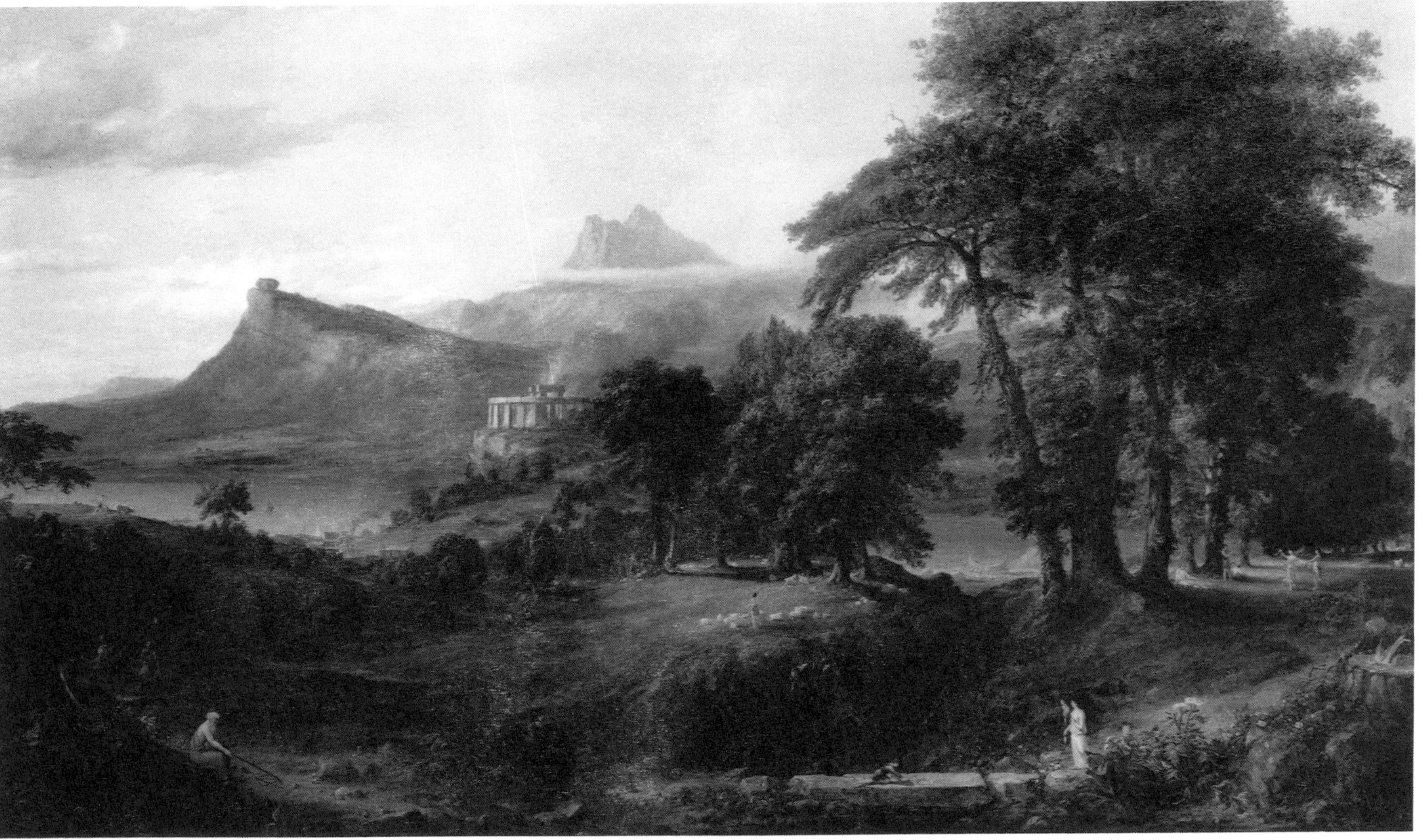

Figure 80. *The Course of Empire: The Pastoral State* (1834); oil on canvas; 100 x 161 cm. Collection of the New-York Historical Society.

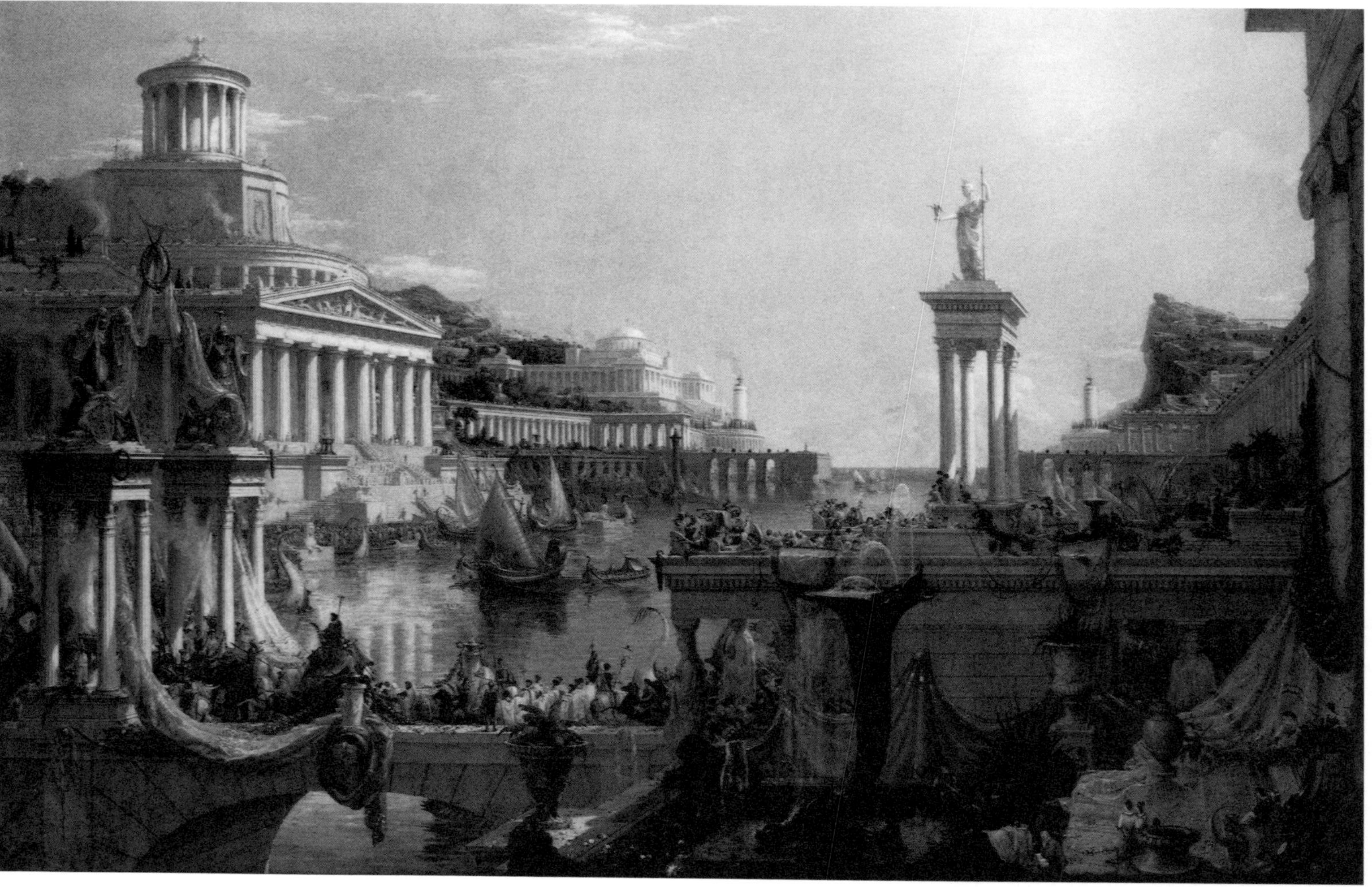

Figure 81. *The Course of Empire: The Consummation of Empire* (1835–36); oil on canvas, 130 x 193 cm. Collection of the New-York Historical Society.

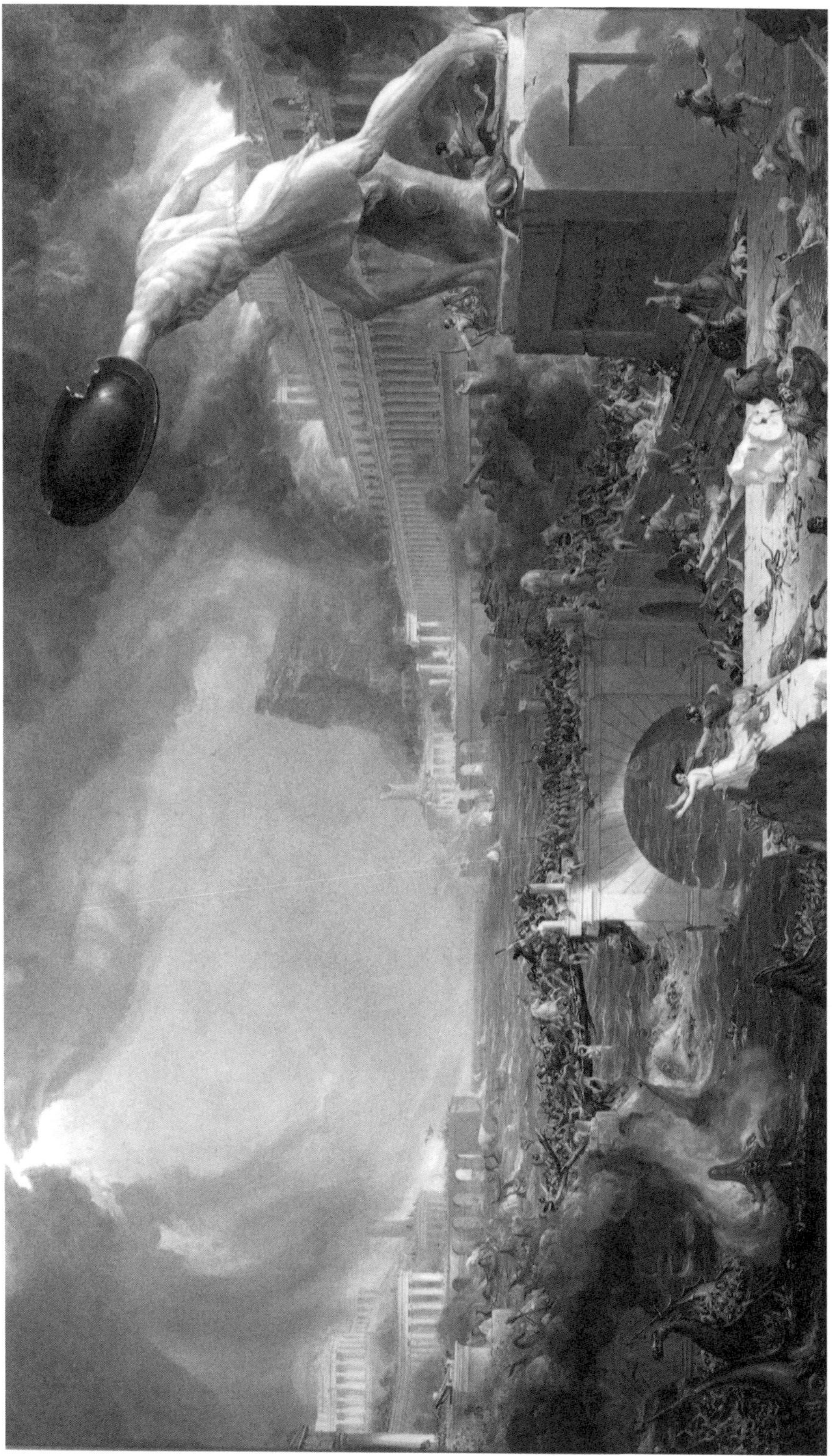

Figure 82. *The Course of Empire: Destruction* (1836), oil on canvas; 100 x 161 cm. Collection of the New-York Historical Society.

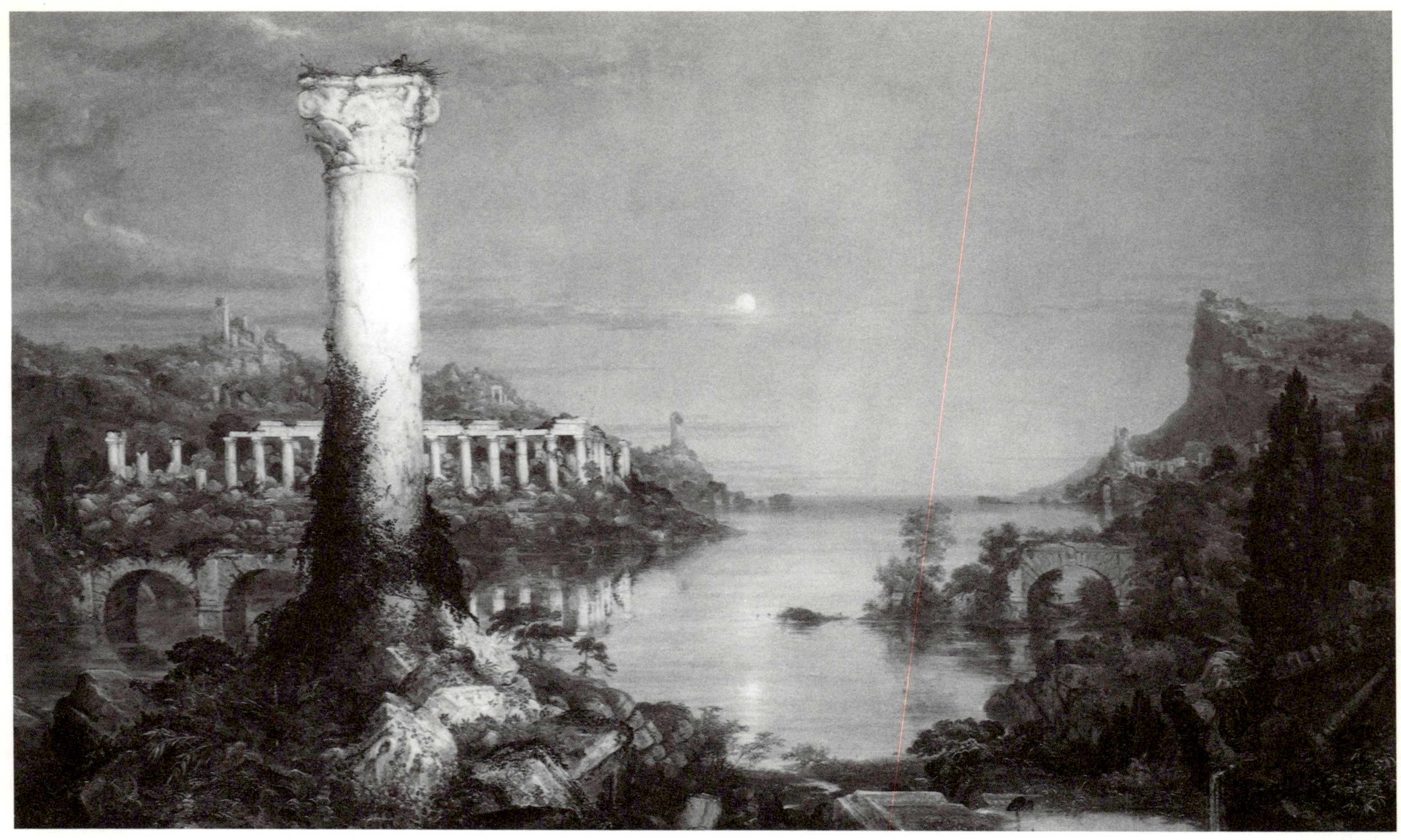

Figure 83. *The Course of Empire: Desolation* (1836); oil on canvas, 100 x 160 cm. Collection of the New-York Historical Society.

ledge in the background of each of the five paintings that make up the series. The paintings trace the rise and fall of a civilization from the *Savage State* (fig. 79), to the *Pastoral State* (fig. 80), to *Consummation* (fig. 81) in which the empire reaches its peak, to *Destruction* (fig. 82) brought on by human weakness—greed, corruption, vanity, and vainglory. In the final painting, *Desolation* (fig. 83), the relics of empire are being slowly absorbed and reclaimed by nature. The rock serves as a fixed reference point for the viewer, indicating the geographical continuity between the scenes. Cole intended the series to express a cyclical vision of history in which, as he explained it, "nations have risen from the savage state to that of power and glory, and then fallen to become extinct."[42] The boulder, sitting imperturbably above each scene, works as a reference to the first civilization that passed through these stages, which was built by the offspring of Adam and Eve and then fell into a moral degeneracy so appalling that God swept it away in the waters of the Flood. The boulder can be read as a remnant of that civilization, a reminder of its watery destruction, and a warning to all those who follow. Its presence on the background escarpment underscores the themes of the series: the cyclical nature of history, the inherent sinfulness of humankind, and the terrible wrath of God.

In *The Notch of the White Mountains (Crawford Notch)* of 1839 (fig. 84), erratic boulders (sitting atop the naked cliff on the left) are again associated with a scene of tragedy and destruction. This New Hampshire valley abounded in frightful associations for the nineteenth-century viewer, as John Sears has pointed out, and geology played a part in these associations.[43] Crawford Notch, it was argued at the time, was a place where the effects of the Deluge were written clearly on the surface topography. The Notch itself was believed to have been excavated by the Flood, while the presence of erratic boulders high on the clifftops and mountainsides attested to the depth and ferocity of the torrential waters.[44] Not only did the landscape bear evidence of this great cataclysm of the past, but it was also the site of a recent tragedy known at the time as the Willey Disaster.

42. Cole to Luman Reed, 18 September 1833; quoted in Wallach, "Cole, Byron, and *The Course of Empire*," 376.

43. My information about Crawford Notch and the Willey Disaster is based largely on John Sears, *Sacred Places: American Tourist Attractions in the Nineteenth Century* (New York, 1989), chap. 4. Sears reproduces Cole's painting of Crawford Notch in this chapter, but he does not discuss it. My interpretation of Nathaniel Hawthorne's fictionalized account of the Willey Disaster is also my own.

44. See, for example, John Carver, *Sketches of New England* (New York, 1842): "We reached the Notch just after noon. The entrance of the chasm is formed by two rocks standing perpendicularly at the distance of twenty-two feet from each other. . . . This opens you into a narrow defile, extending two miles in length, between two huge cliffs, apparently rent asunder by some great convulsion of nature. This convulsion, Dr. Dwight thinks, was that of the deluge, since there are no proofs of volcanic action anywhere in this region" (pp. 90–91).

On the night of 28 August 1826, an avalanche of rocks and mud roared down one of the sloping walls of the Notch and buried alive Samuel Willey and his wife, their five children, and two hired men. The family had experienced avalanches previously, and only two months earlier a rock slide had come very close to their home. Willey then constructed a shelter away from the house, and when they were awakened by the rumble of falling rocks on the night of 28 August, it was to this refuge that they fled—right into the path of the oncoming slide. Just before the avalanche reached the house, it changed course, leaving the building untouched. If the Willeys had stayed in their beds, they would have been spared.

The Willey Disaster, as Sears has observed, posed to Cole's contemporaries an interpretive problem in terms of divine order. The diluvial associations of the spot naturally suggested divine retribution, and parallels between the Deluge and the Willey Disaster were drawn at the time. Silliman, in a communication to the *American Journal of Science* in 1829, noted that the Willey catastrophe "enables one to form some feeble conception of the universal effects of the vindictive deluge which once swept every mountain and ravaged every plain and defile."[45] In this context, the Willey catastrophe could be interpreted as a warning, a reminder of the wages of sin. But by every account the Willeys were a humble, upright, and industrious family. No hint of immorality, no taint of scandal ever touched them. They lived quietly, tending their small farm, catering to the needs of travelers who passed through the Notch. Seeking explanations, most interpreters were forced to fall back on the mysterious workings of God.

Cole's representation of the site is itself somewhat mysterious. It may be a topographic view of Crawford Notch as it appeared to Cole on his visit to the site in 1839. Comparing the painting to a drawing of the Notch that Cole made on that trip (fig. 85) reveals a close correspondence, even though in the painting Cole has shifted the point of view to the left. The configuration of the topography (the shape of the notch and the mountains) is consistent, as are the dead trees and the erratic boulders on the cliff. To that extent Cole remained true to what he saw on the spot. Yet he has made enough changes and additions to the scene to suggest that this is not a simple topographic view but rather a narrative picture, a reconstruction describing the site in the hours before the Willey Disaster. For example, the foliage has been returned to the trees around the house, and the dwelling is clearly occupied. Smoke curls from the chimney, and a man and child have emerged from the house to greet an

---

45. Quoted in Sears, *Sacred Places*, 80.

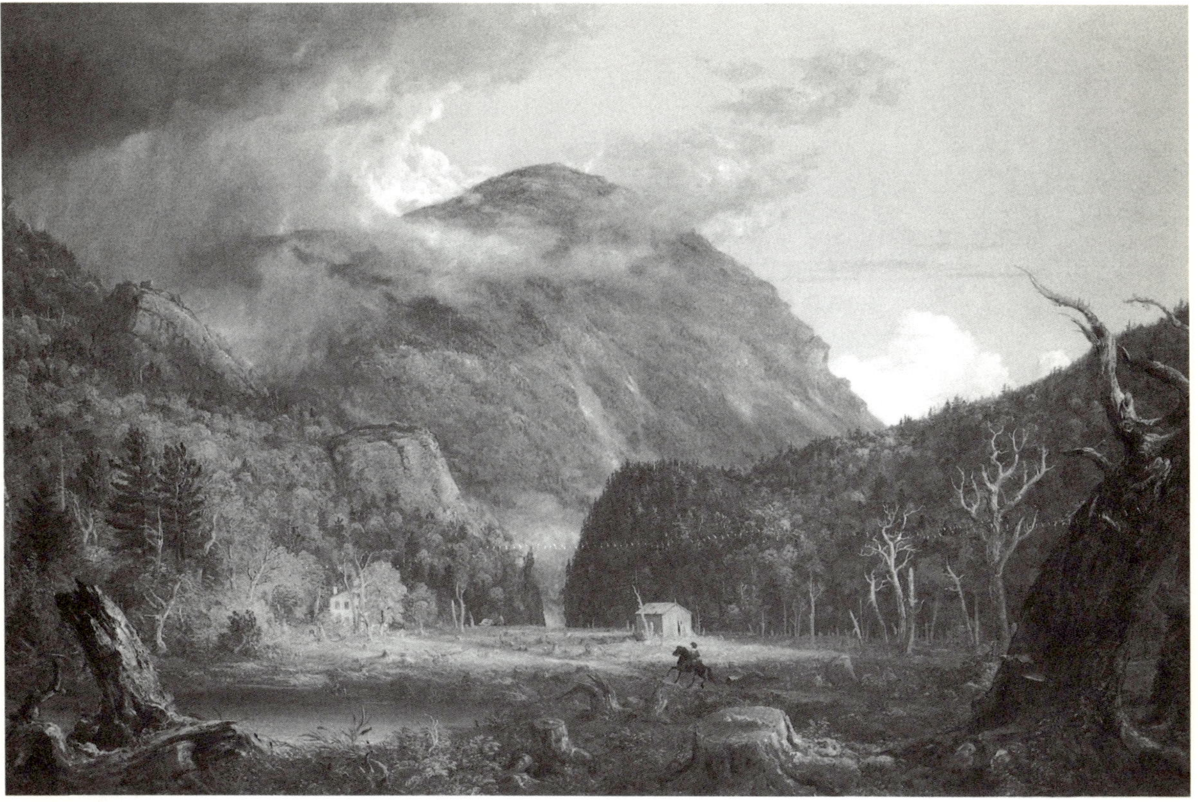

Figure 84. *The Notch of the White Mountains (Crawford Notch)* (1839); 101 x 156.2 cm. National Gallery of Art, Washington, D.C., Andrew W. Mellon Fund.

approaching visitor. Goats graze beside the shed, and a stage coach has just passed along the road.

Despite these references to domestic felicity, there are intimations of the approaching tragedy. The horse in the foreground seems to be rearing back as if frightened. Huge black storm clouds unleash torrents of rain on the mountain peaks, while on the background mountain several diagonal streaks of swirling brown pigment suggest avalanches of mud and stone tumbling into the Notch. The dead trees add a spectral presence, their twisted branches evoking the agony of death. Even the light seems to allude to the tragedy. A narrow band of golden light bathes the house and a section of the yard, demarking that slim zone of safety in which the Willeys might have found salvation.

Yet if the painting is a historical view of the Notch on the eve of the Willey Disaster, why is the traveler given such prominence? This might be explained if we read the painting as referring not to the Willey Disaster alone but also,

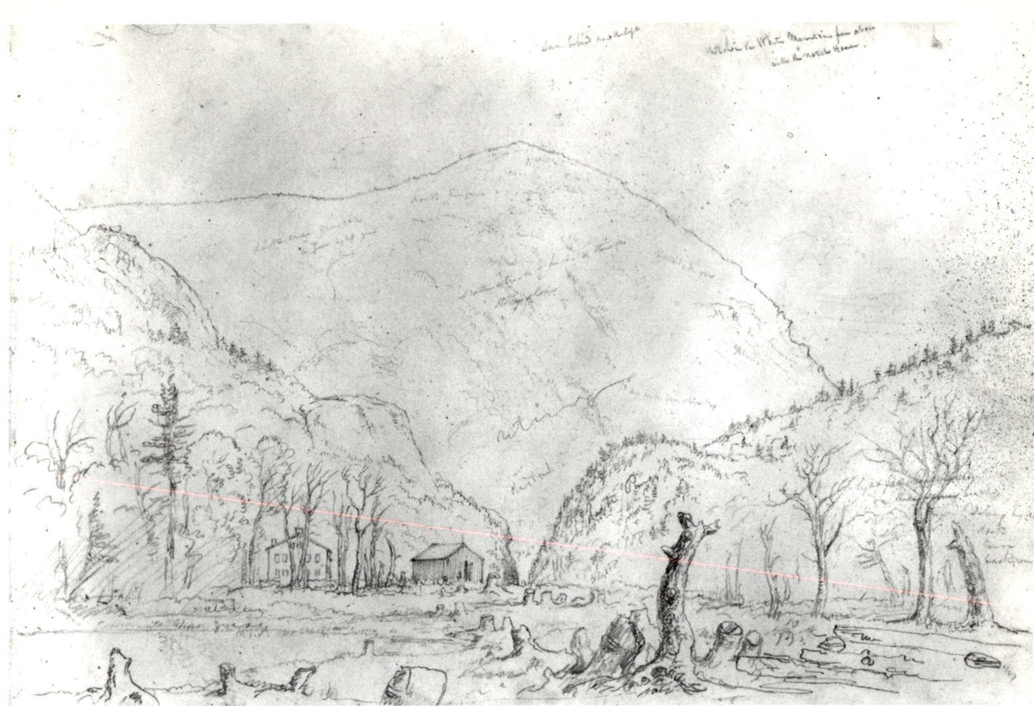

Figure 85. *Notch in the White Mountains, from Above with the Notch House* (1839); pencil on paper, 28.3 x 42.8 cm. Art Museum, Princeton University, Frank Jewett Mather Jr. Collection.

more particularly, to Nathaniel Hawthorne's fictional account of it—a short story entitled "The Ambitious Guest," published in his *Twice-Told Tales* of 1837.[46] If this was Cole's intention then the traveler is not a random tourist passing through the Notch but the central figure in Hawthorne's short story.

Hawthorne's story takes place on the evening of the disaster. The Willey family is seated around the fire basking in domestic bliss when a young man appears at their door seeking shelter for the night. They absorb him into their family circle and listen attentively as he spills out to them his hopes for the future, his desire for fame. As he speaks, the exemplary Willeys begin to feel stirrings of ambition in their own breasts. Mr. Willey dreams aloud of becom-

46. *Twice-Told Tales* was Hawthorne's first successful volume, and was published at least partially through the efforts of Cole's Boston patron Samuel Griswold Goodrich, who liked to think of Hawthorne as his protegé. (In fact, in the late 1820s and early 1830s, Goodrich drew on the talents of both Cole and Hawthorne to fill his publications, including *The Token* and *New England Magazine*.) *Twice-Told Tales* was well received, garnering positive reviews in a number of journals read by Cole, including *The Knickerbocker* and the *North American Review*. The latter review, written by Henry Wadsworth Longfellow, hailed Hawthorne as a man of genius and an American original. See Arlin Turner, *Nathaniel Hawthorne: A Biography* (New York, 1980), chaps. 7 and 8. On Cole's relationship with Goodrich, see Parry, *Art of Thomas Cole*, 37, 54, 85–86.

ing a country squire and his daughter's thoughts turn to suitors. When the elderly grandmother begins to dwell on thoughts of her approaching death, a superstitious and rather vain preoccupation takes hold of her: she remembers being told that if anything is amiss with a corpse it will not be able to rest easily in its grave. She then begs her family to hold a mirror over her face as she lies in her coffin so that she will be able to look at herself and see that all is right. It is at just this moment, when the Willeys' thoughts have been turned toward death, superstition, and ambition, that the avalanche claims their lives. In Hawthorne's story, the concluding irony is that the young man so intent on winning earthly renown dies without leaving a trace, his presence unsuspected by those who sifted through the slide for remains.

Cole's painting is clearly not a literal illustration of Hawthorne's story. The tale takes place entirely within the Willeys' abode, while Cole presents an exterior scene. Hawthorne's setting is nighttime, Cole's daytime. Hawthorne's young man is described as a pedestrian; Cole's is an equestrian. Yet these departures from the story are within the bounds of artistic license, adjustments made to the narrative to better suit Cole's own gifts. By shifting the setting he could play to his strength as a landscape painter; by changing the hours he could avoid the difficulties of a nocturnal scene, and by placing the figure on horseback Cole could give him a much more prominent place in the composition, in keeping with his central role in the story.

Hawthorne's tale, with its emphasis on the consequences of ambition, would I think have had strong personal resonance for Cole, who felt that ambition was his own "master sin." According to Noble, both his biographer and his minister, Cole attributed "the most serious of [his own religious trials] to ambition. To win renown had been from his youth a ruling passion. In the light of religion that passion stood forth as his master sin. To vanquish it his most difficult labor."[47] By alluding to Hawthorne's story in *The Notch of the White Mountains*, Cole could embed a strong moral in the canvas, a moral much like that propounded in his *Course of Empire* series. In *The Course of Empire,* it is pride and ambition (among other sins) that bring about the fall of the empire. Hawthorne's story points to the same lesson in a more individual context, the idea that pride and ambition precede a fall.

If this painting alludes to Hawthorne's story, then the sin of ambition may be offered here as the solution to the puzzle of the Willeys' demise. But this is a rich and complicated picture, and through the details of the painting Cole

---

47. Noble, *Life and Works of Thomas Cole*, 301–2. See also Kenneth James LaBudde, *The Mind of Thomas Cole* (Ph.D. diss., University of Minnesota, 1954), 192–93.

hints at another possible solution. Prominently displayed in the foreground of the picture are two stumps, their tops tilted toward us to display the marks of the axe.[48] Dozens more stumps punctuate the middle ground. At the time Cole painted this picture, he had become deeply disturbed by the devastation of the American wilderness. In his prose, poetry, and paintings he lamented the ravages that the axe was daily inflicting on nature in America. Here he draws a connection between the violence man has wrought on nature and the violence nature has visited on man. In this context, the Willey Disaster could be read as retribution for the desecration of what Cole's friend William Cullen Bryant called "God's first temples."[49]

In selecting Crawford Notch for his subject, Cole could rely on the site's rich scientific, historical, and literary associations to carry the viewer's mind beyond this specific locale to ponder the moral implications of the scene: the frailties of humankind, the brevity of life, and the need to be prepared for death. At the same time, the picture announced the potential of landscape painting to rival history painting in the complexity and profundity of its themes.

In *The Notch in the White Mountains,* as in so many of Cole's landscapes, geological ideas are woven into a dense fabric of varied associations and meanings. Neither here nor in Cole's other works did geology provide the main motive for the painting, yet again and again he was able to draw upon geology to reinforce and bolster his depiction of religious and moral themes. At the same time, his geological knowledge helped him forge ties to his upper-crust patrons, providing one of the chief means of establishing himself in his own eyes and in theirs as a "gentleman."

Wellesley College

---

48. Baigell, in his brief discussion of this painting in *Thomas Cole,* mentions the stumps and describes them as showing "the violence man has done to nature" (p. 60). For a more general discussion of this feature, see Nicolai Cikovsky Jr., "'The Ravages of the Axe': The Meaning of the Tree Stump in Nineteenth-Century American Art," *Art Bulletin* 61 (December 1979): 611–26; and Novak, *Nature and Culture,* chap. 8.
49. Novak, *Nature and Culture,* 157.

Amy R. W. Meyers is curator of American Art at the Huntington Library, Art Collections, and Botanical Gardens. Her longstanding interest in the relationship between art and science in America began with the writing of her dissertation, "Sketches from the Wilderness: Changing Conceptions of Nature in American Natural History Illustration, 1680–1880" (Yale University, 1985). She has contributed essays to *Views and Visions: American Landscape before 1830* (1986); *John James Audubon: The Watercolors for "The Birds of America"* (1993); and *Mark Catesby's "Natural History of America": The Watercolors from the Royal Library, Windsor Castle* (1997). She is also co-editor, with Margaret Pritchard, of *Empire's Nature: Mark Catesby's New World Vision,* a volume of essays forthcoming from the Colonial Williamsburg Foundation and the Omohundro Institute of Early American History and Culture. She is presently organizing a traveling exhibition on Philadelphia natural history illustration from 1740 to 1840, which will open at the Huntington in the winter of 2002.

Rebecca Bedell teaches in the art department at Wellesley College. She is currently completing a book on geology and American landscape painting, 1825–75.

David R. Brigham, curator of American art at the Worcester Art Museum, is currently preparing an on-line catalogue of its early American paintings, and he recently organized an exhibition on American impressionism. His research on the subscribers to Mark Catesby's *Natural History of Carolina* will be published in *Empire's Nature: Mark Catesby's New World Vision.*

Kenneth Haltman teaches history of art and American studies at Michigan State University. His recent publications include "Titian Ramsay Peale: Specimen Portraiture, or Natural History as Family History," in *The Peale Family: Creation of a Legacy* (exhibition catalogue, 1997); and "Antipastoralism in Early Winslow Homer," in the *Art Bulletin*. His book *Figures in a Western Landscape: Expeditionary Art and Science in the Early Republic* is forthcoming from Princeton University Press.

Therese O'Malley is associate dean at the Center for Advanced Study in the Visual Arts at the National Gallery of Art. She is currently working on a reference work, "Keywords in American Landscape Design," and a book on the botanic garden in America from the colonial period through the twentieth century.

Linda Dugan Partridge, who teaches at Marywood University, is currently a National Endowment for the Humanities research fellow at the Winterthur Library. She is completing work on a study of John James Audubon.

Laura Rigal teaches English and American Studies at the University of Iowa. Her essays have appeared in *Theatre Journal* and in *American Iconology: New Approaches to Nineteenth-Century Art and Literature,* edited by David Miller (1995). Her book *The American Manufactory: Art, Labor, and the World of Things in the Extended Republic, 1783–1828* is forthcoming from Princeton University Press.